The Accelerating Angle Curve

The Accelerating Angle Curve
Adventures in Symmetry

Stephen P. Hershey

VERTILAB

Published by VERTILAB
Sarasota, Florida, USA
www.vertilab.com

ISBN 978-0-9966971-1-8
Library of Congress Control Number: 2017916150

Typeset in LaTeX
Copyeditor: Sue Hargis Spigel
Front cover figure: AAC [360: 0 83 233 214]

Spirograph® is a registered trademark of Hasbro, Inc.

First edition: December 2017

*Dedicated to my mother (the artist) and
my father (the engineer)*

I have no private Interest in the Reception of my Inventions by the World, having never made nor proposed to make the least Profit by any of them.

— Benjamin Franklin, *Letter to Lebègue de Presle* (1777)

Contents

Preface

In 1978, I was a 15-year-old boy exploring the graphics capability of a newfangled machine called a personal computer — the Apple II. I invented an algorithm for generating two-dimensional curves that I called *accelerating angle*. I noticed what appeared to be a surprising correlation between curve order and curve rotational symmetry.

In 1986, I was in graduate school in electrical engineering at Purdue, and as a hobby project ported the accelerating angle curve generation algorithm from the Apple II to the Macintosh in order to take advantage of the Macintosh's superior graphics capability. However, I did little further exploration of accelerating angle curves.

For the next 30 years, accelerating angle curves sat in a back corner of my mind, gently nagging me with the question, "Why is there a correlation between curve order and curve rotational symmetry?"

Finally, in 2016 I began studying accelerating angle curves in earnest, completing my studies in 2017. I quickly discovered that there is no simple correlation between curve order and curve rotational symmetry.

This book is a gift from my 55-year-old self to my 15-year-old self, saying, "*This is what you stumbled upon so long ago!*"

Stephen P. Hershey
Sarasota, Florida, USA
December 2017

List of Figures

List of Tables

1 Introduction

> I aim at two things: on the one hand to display the great variety of applications of the principle of symmetry in the arts [and] nature, [and] on the other hand to clarify step by step the philosophico-mathematical significance of the idea of symmetry.
>
> — Hermann Weyl, *Symmetry* (1952)

1.1 Overview

Simple recurrence relations that generate complex results have been objects of fascination for many years, especially since advances in computer and display technology have eliminated the tedium of generating such results manually. The accelerating angle algorithm can generate planar, segmented curves of surprising elegance and complexity. Accelerating angle curves (AACs) are interesting both for their aesthetics and as a subject for analysis. This book addresses AAC aesthetics by providing a gallery of AACs. The analytical questions considered by the book deal primarily with the rotational and reflectional symmetries of closed AACs.

The four algorithms presented in this book — AAC generation, predicting the number of AAC rotational symmetries, determining the number of AAC rotational symmetries, and determining the number of AAC reflectional symmetries — have been reduced to practice and thoroughly tested. The AAC generation algorithm employs trigonometric functions, so AAC point data often approximate irrational values. In the interest of simplicity, all AAC point data presented in this book are rounded to the nearest thousandth place value and the algorithms presented do not compensate for potential inequality due only to roundoff error when checking two numbers for equality. Computer-generated text is presented in a `fixed-width` `font`. The subscripted angle a_n appears as the indexed angle a[n] in computer-generated text. The modulus function employed throughout this book always returns a nonnegative integer.

1

All of the AACs in this book were generated by the author. Techniques for exploring the creative potential of the AAC generation algorithm are beyond the scope of this book.

1.2 Organization

This book can be organized conceptually into four categories: AAC aesthetics (chapter 2), AAC analysis (chapters 3 – 7), AAC affiliated subjects (chapter 8), and AAC open questions (chapter 9). The book contains a total of 10 chapters:

- Chapter 2 provides a gallery of AACs.

- Chapter 3 describes the AAC generation algorithm.

- Chapter 4 describes the approach to AAC symmetry employed by this book.

- Chapter 5 presents an algorithm that predicts the number of rotational symmetries of a closed AAC given only its generation parameters.

- Chapter 6 presents an algorithm that determines the number of rotational symmetries of a closed AAC given only its point data.

- Chapter 7 presents an algorithm that determines the number of reflectional symmetries of a closed AAC given only its point data.

- Chapter 8 explores the relationship between order 1 AACs and Spirograph hypotrochoid curves.

- Chapter 9 presents some open questions regarding AACs.

- Chapter 10 provides concluding remarks.

A list of abbreviations appears at the end of the book.

2 Accelerating Angle Curve Gallery

This chapter provides 60 examples of closed AACs, selected either for aesthetic value or for variety. Some figures appear to consist of multiple superimposed curves, but this is an illusion — all figures consist of a single curve. The caption for each curve is its *signature*, described in section 3.5 on page 76.

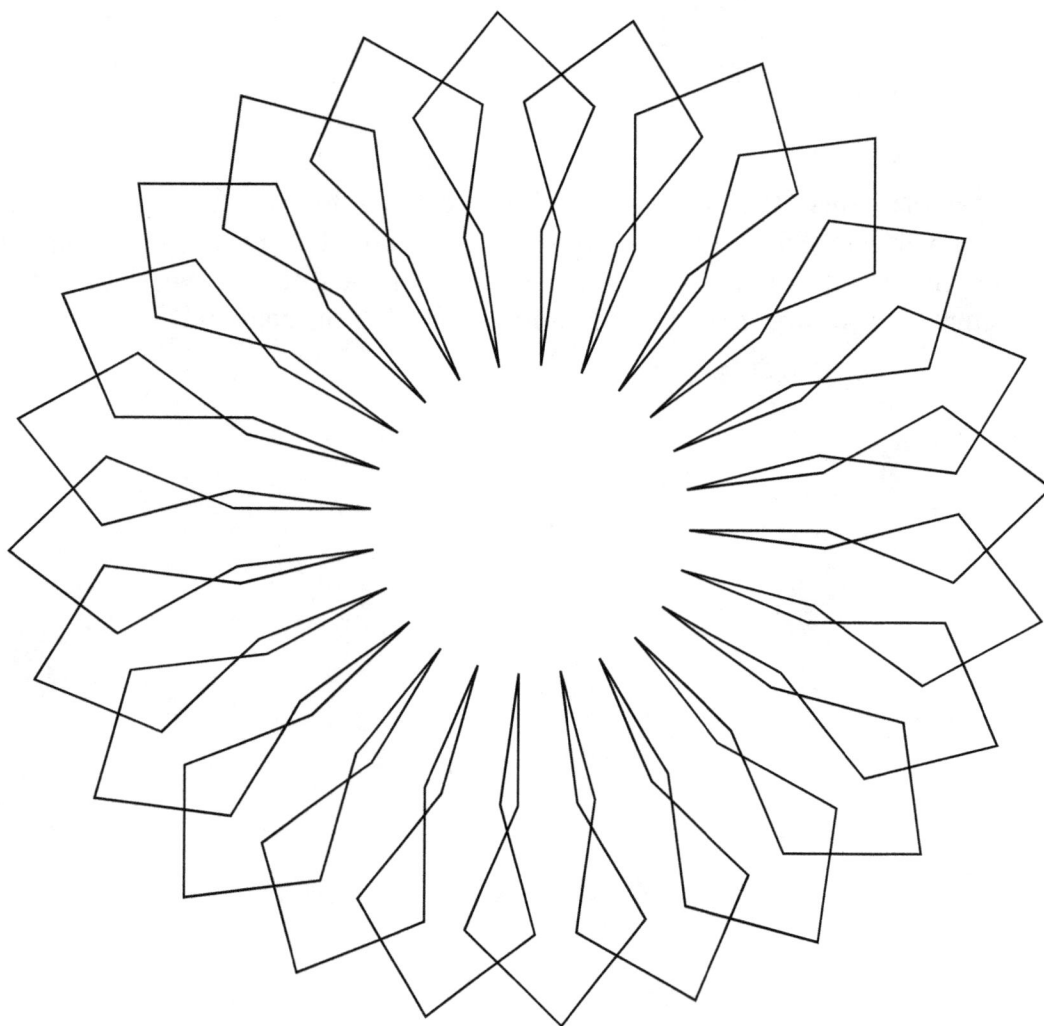

Figure 2.1   AAC [96: 0 90 24 80]

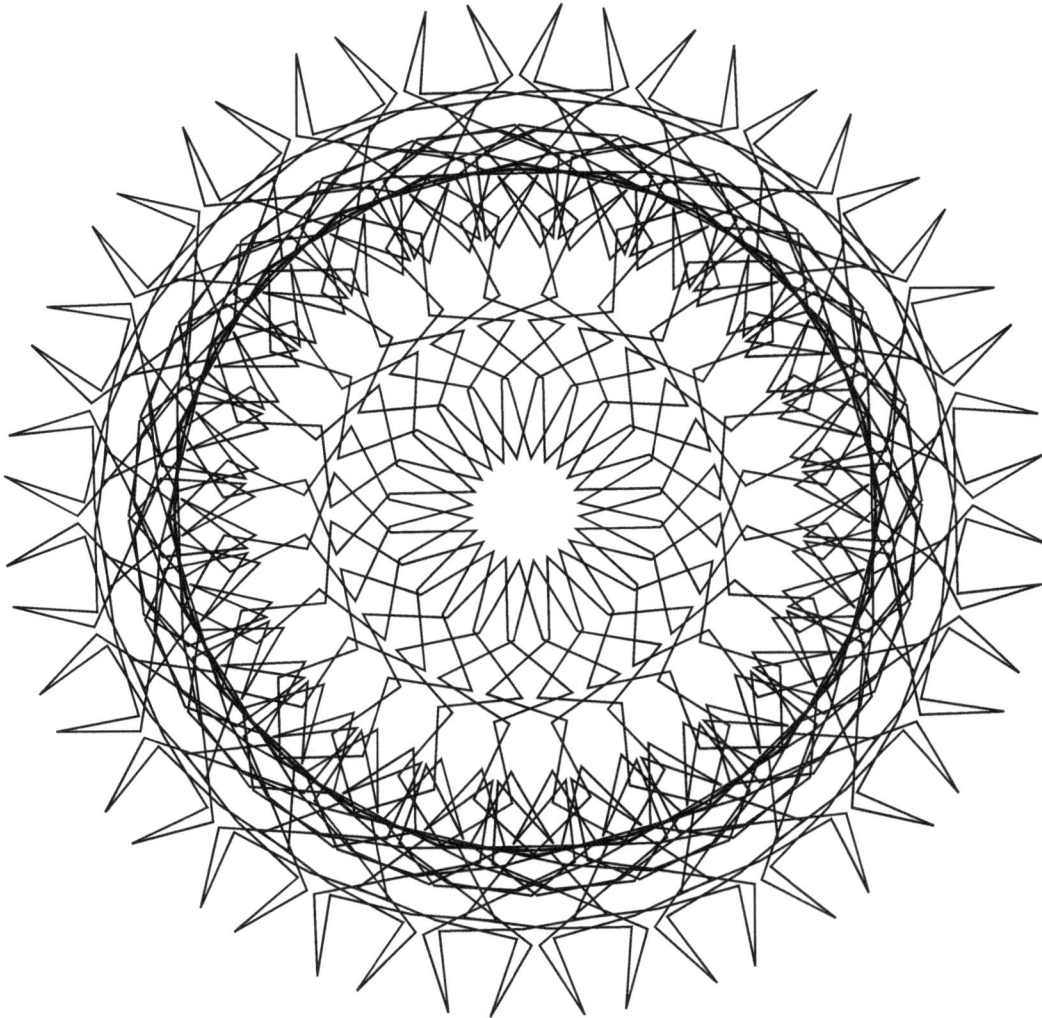

Figure 2.2 AAC [360: 0 1 0 8]

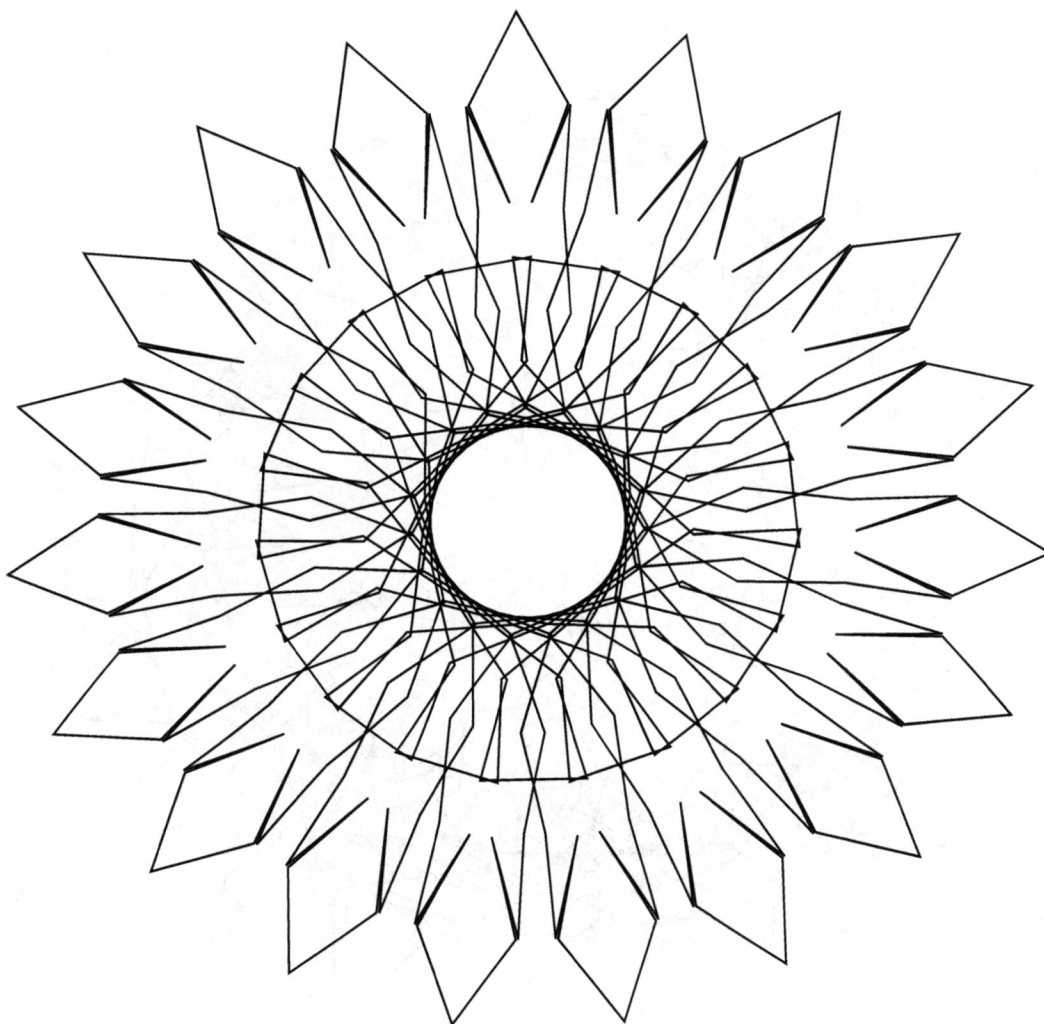

Figure 2.3   AAC [361: 0 46 304 152]

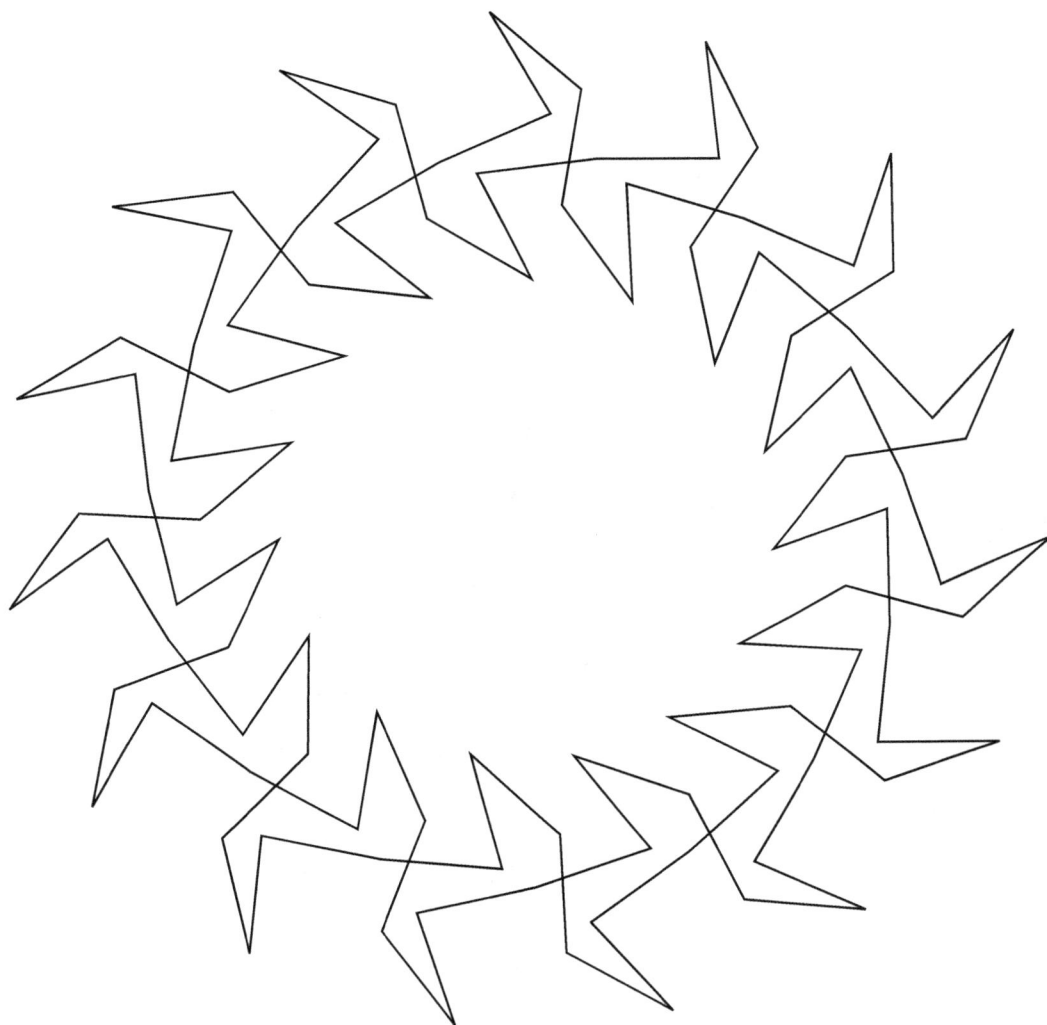

Figure 2.4   AAC [105: 0 28 30]

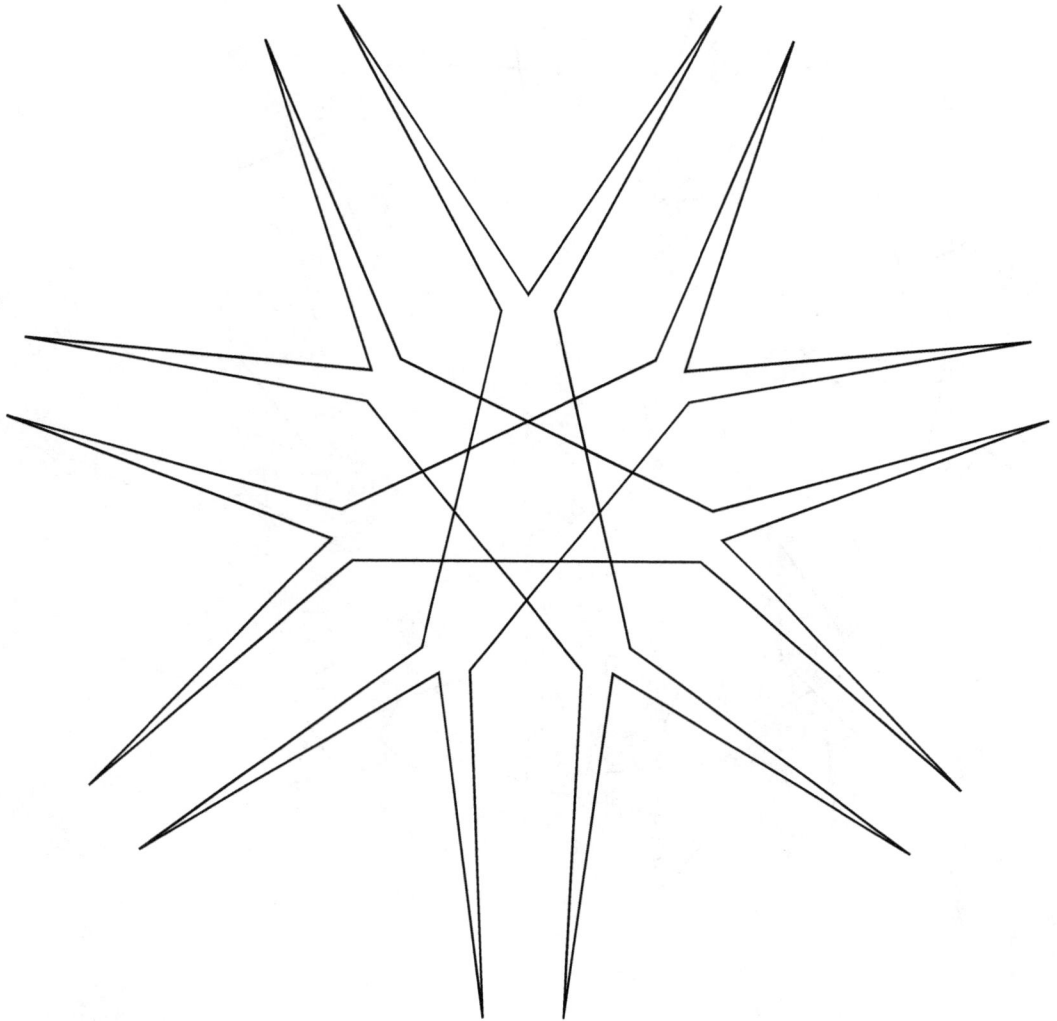

Figure 2.5   AAC [35: 2 17 14 21]

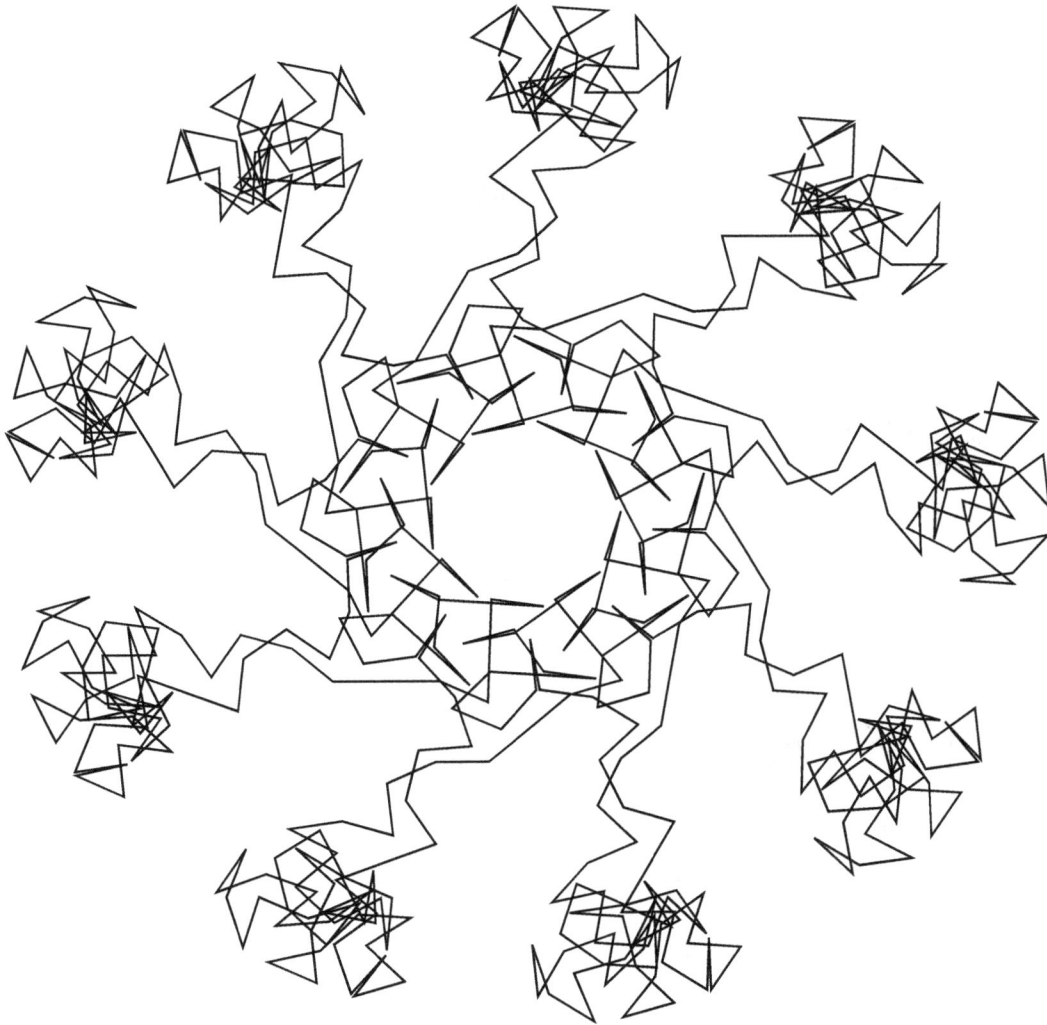

Figure 2.6 AAC [360: 0 0 120 16 24 168 144]

Figure 2.7   AAC [360: 0 0 0 1]

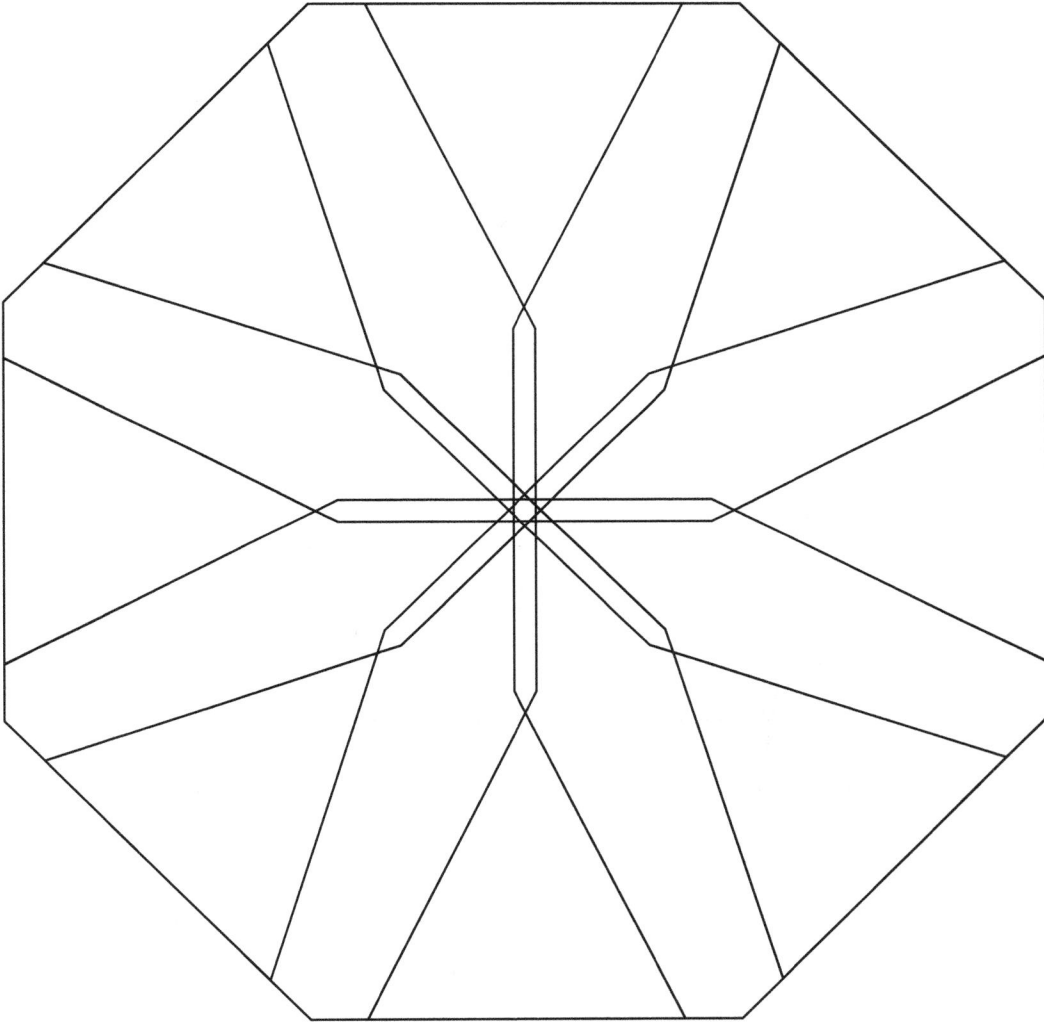

Figure 2.8   AAC [120: 0 9 72 72]

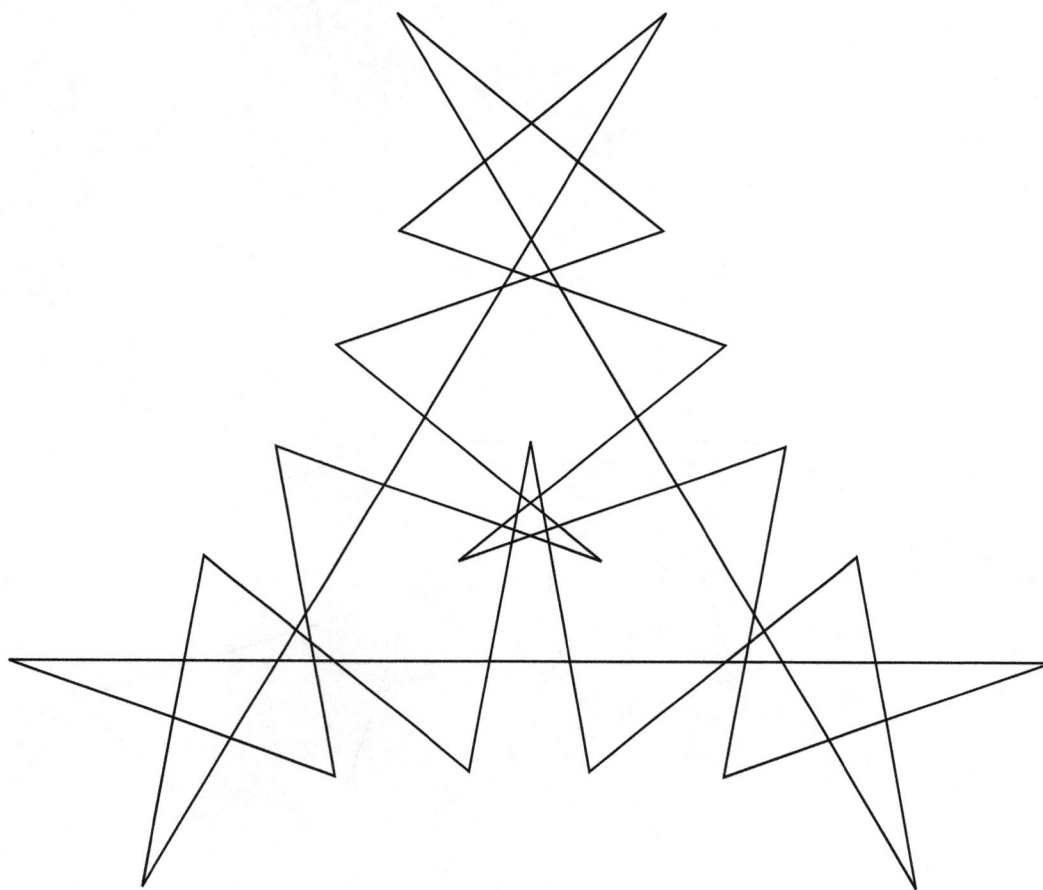

Figure 2.9 AAC [576: 64 192 192 320]

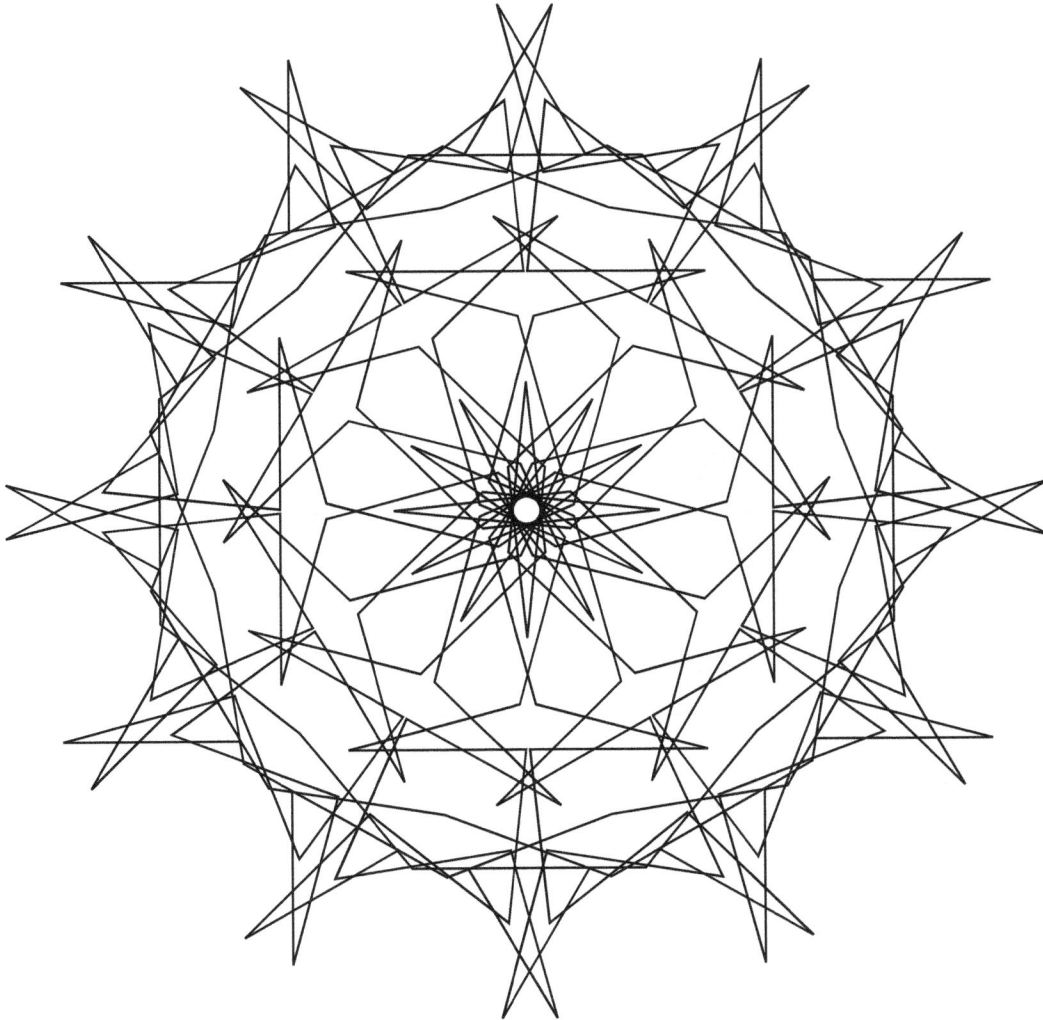

Figure 2.10   AAC [840: 0 280 441 546]

13

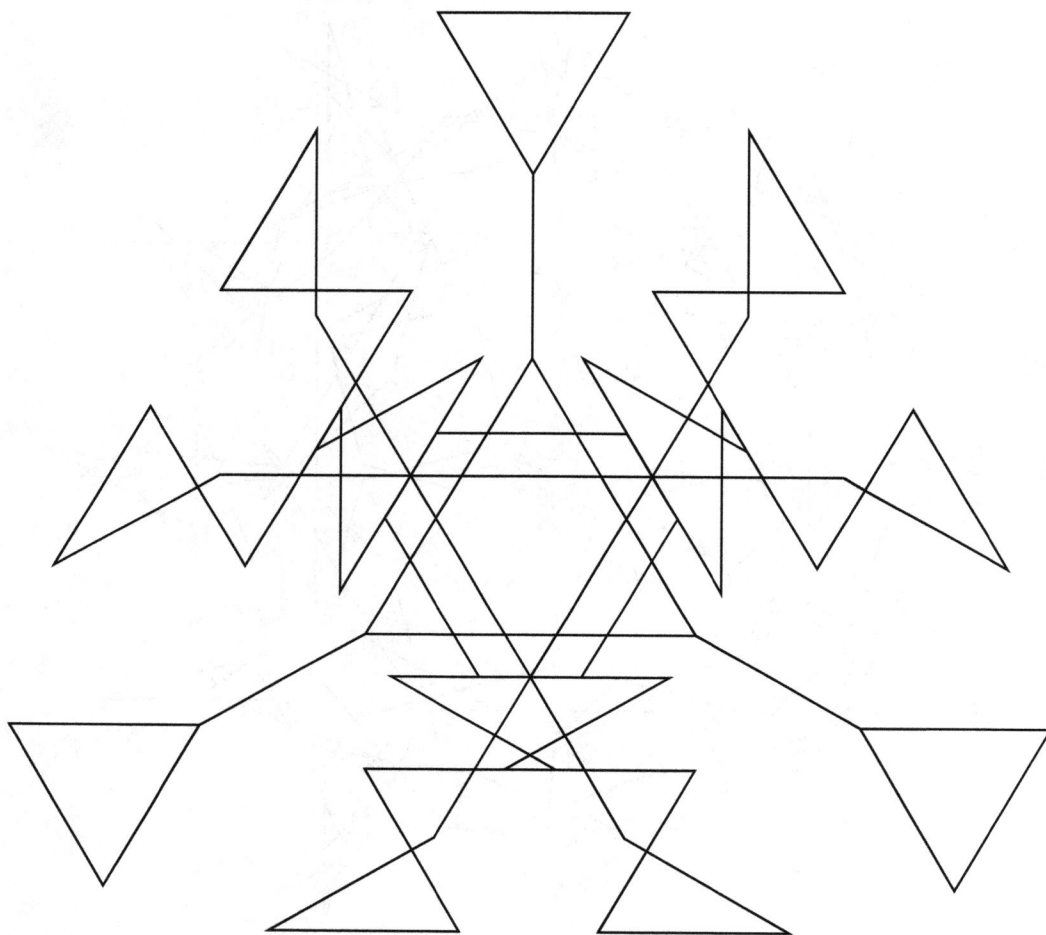

Figure 2.11   AAC [12: 6 10 0 7]

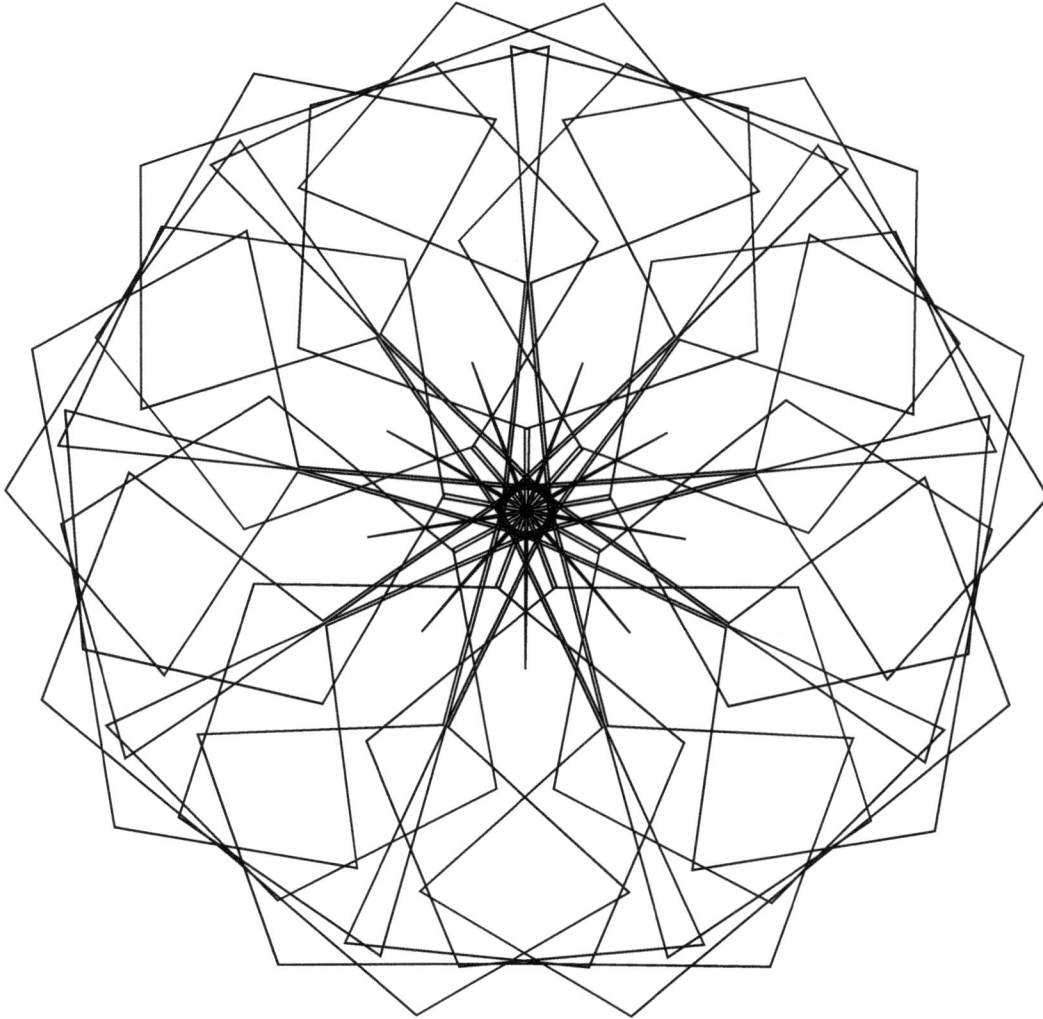

Figure 2.12   AAC [720: 18 142 576 288 360]

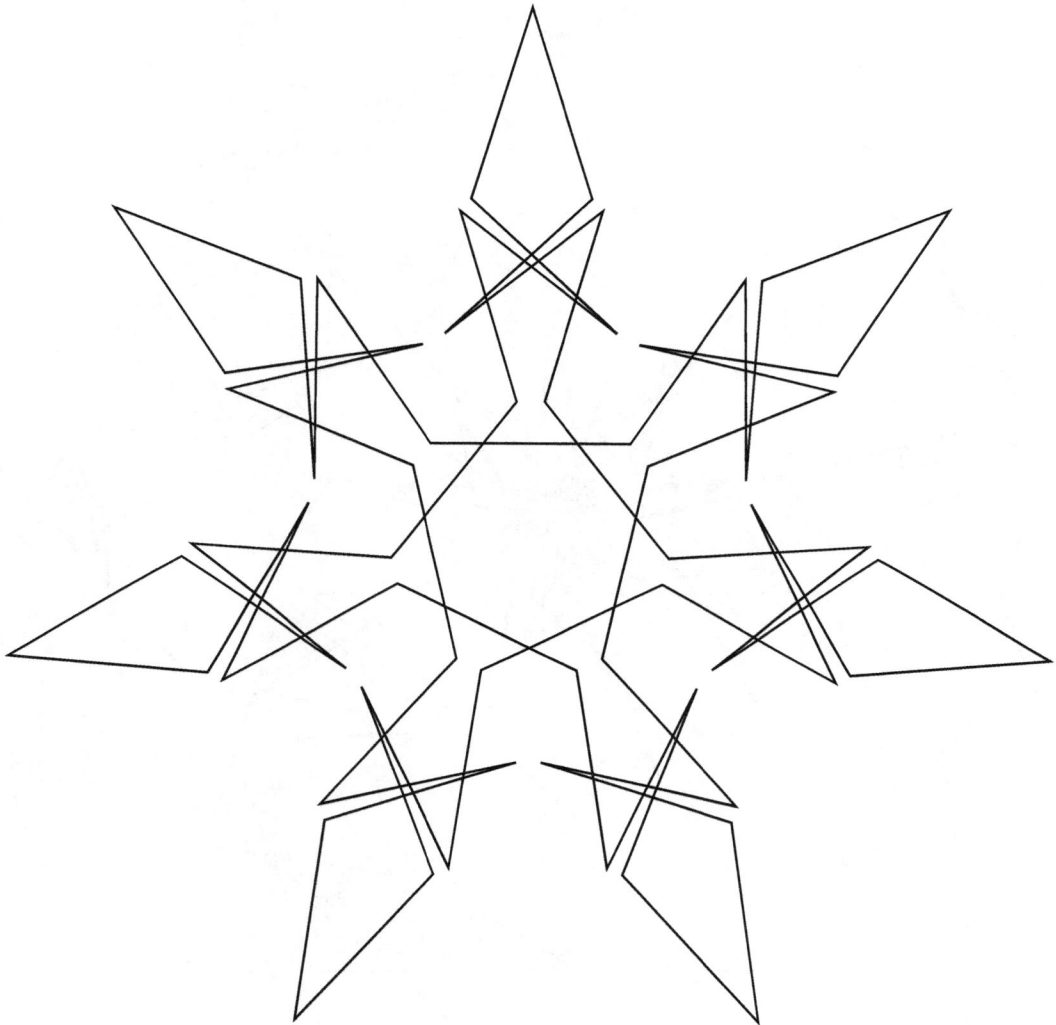

Figure 2.13   AAC [378: 45 68 84 210]

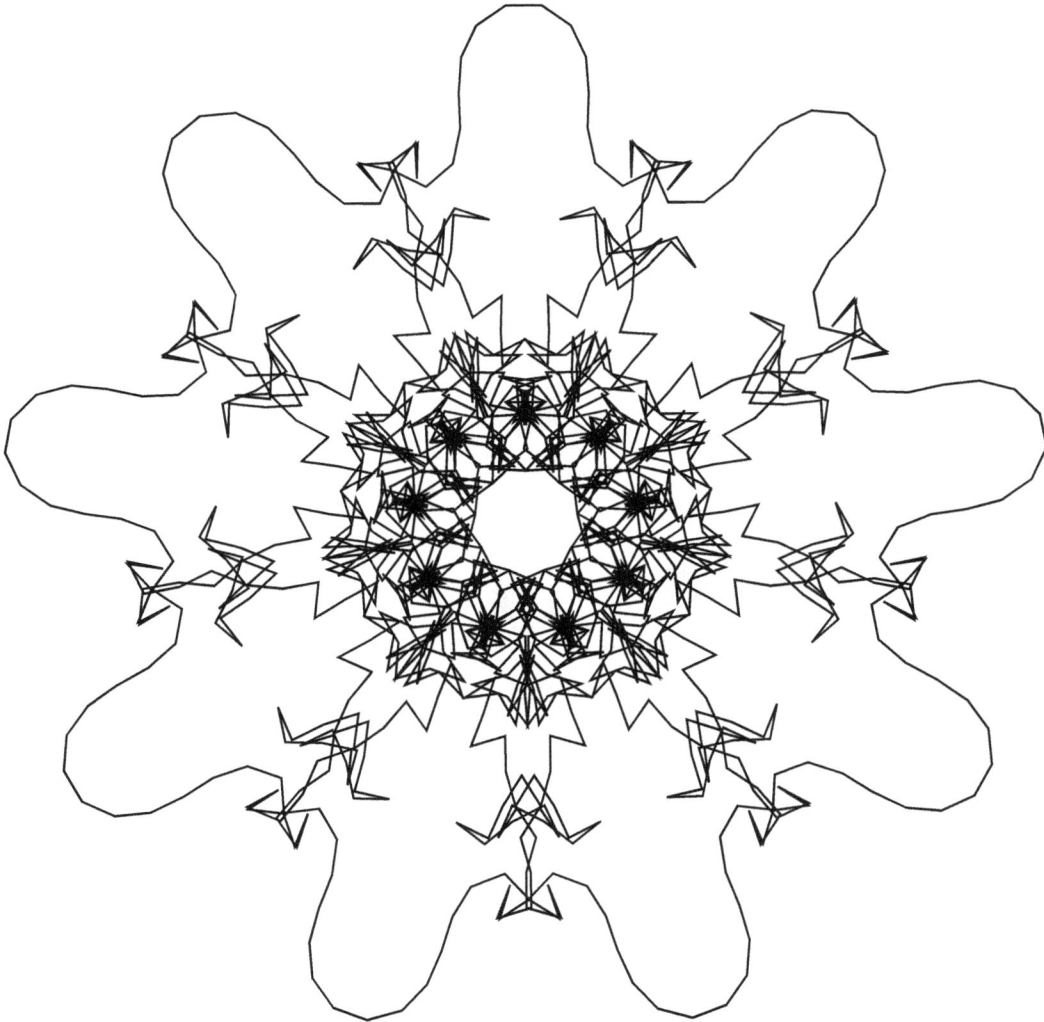

Figure 2.14   AAC [666: 12 187 405 9]

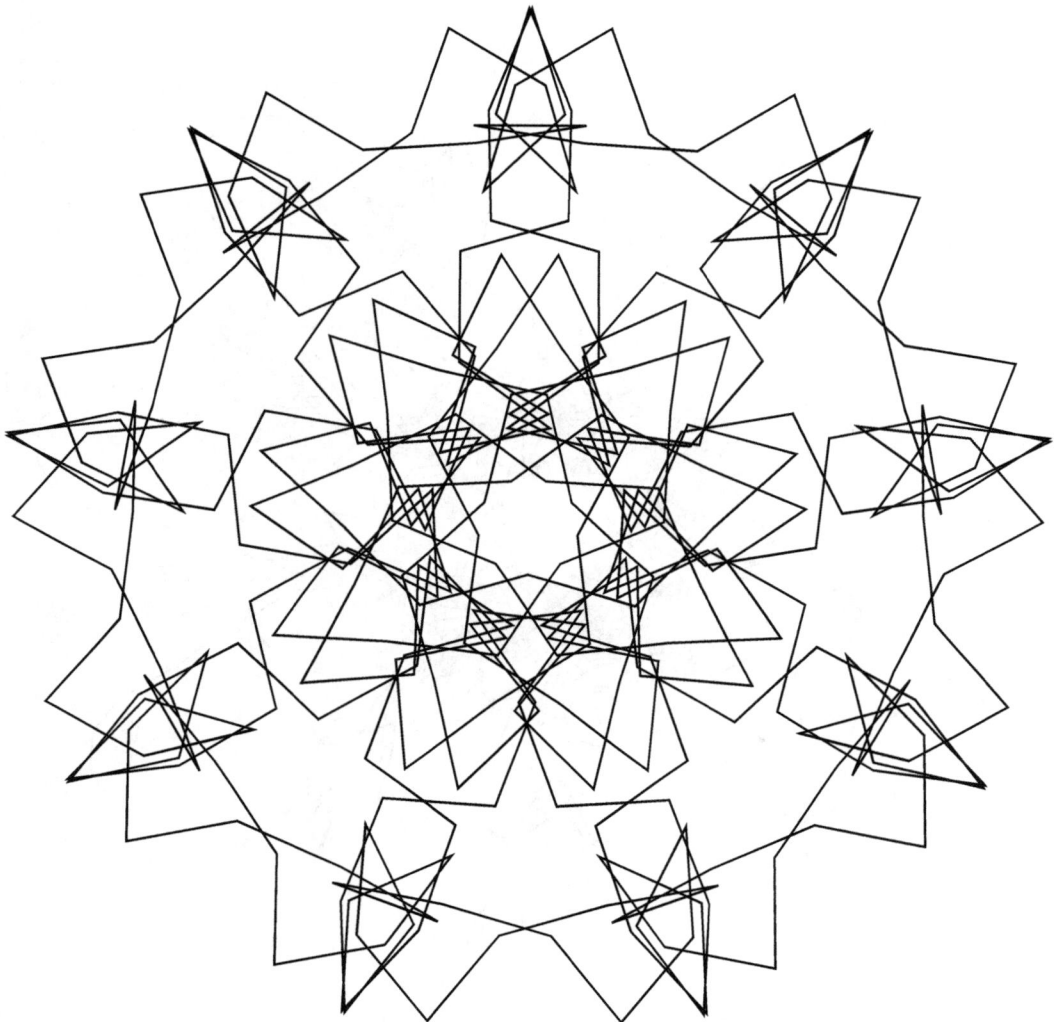

Figure 2.15   AAC [666: 18 530 270 306]

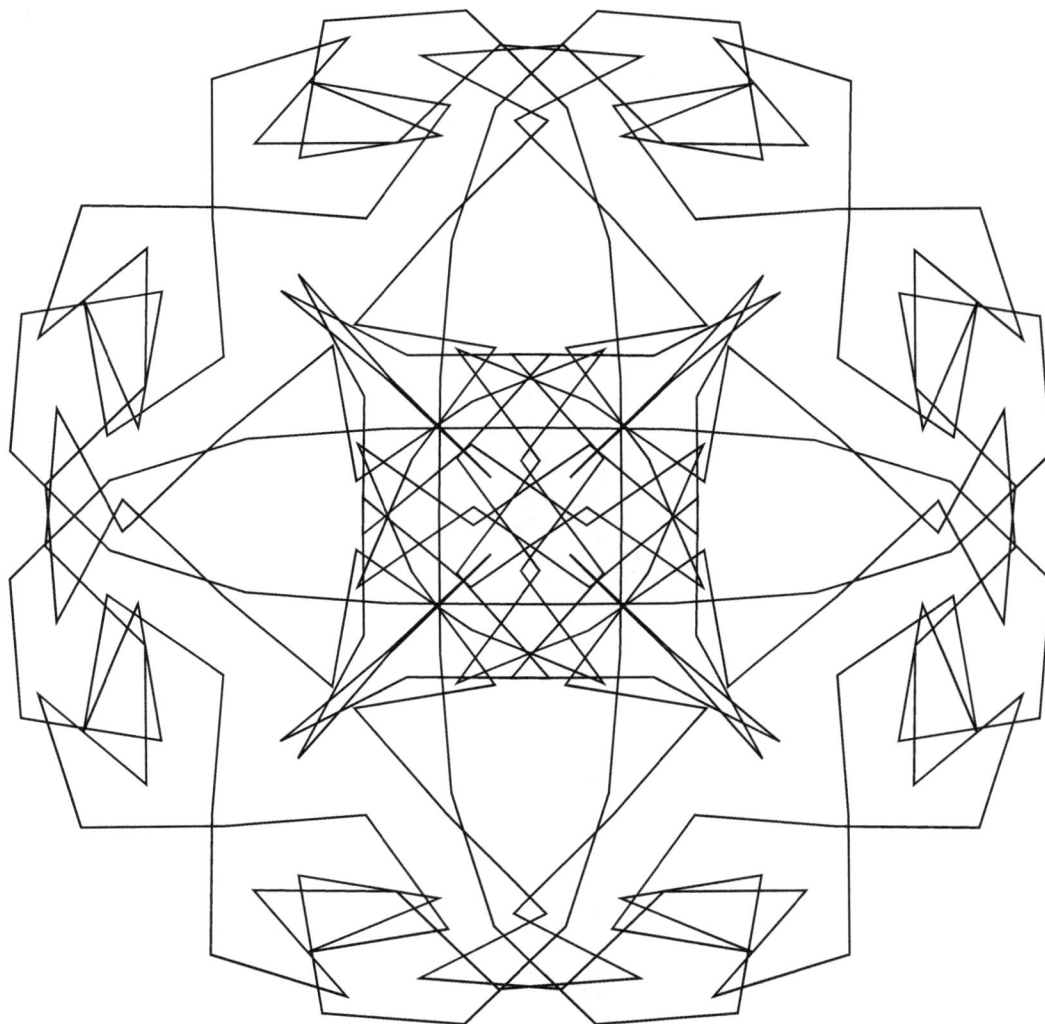

Figure 2.16 AAC [360: 0 4 9 6]

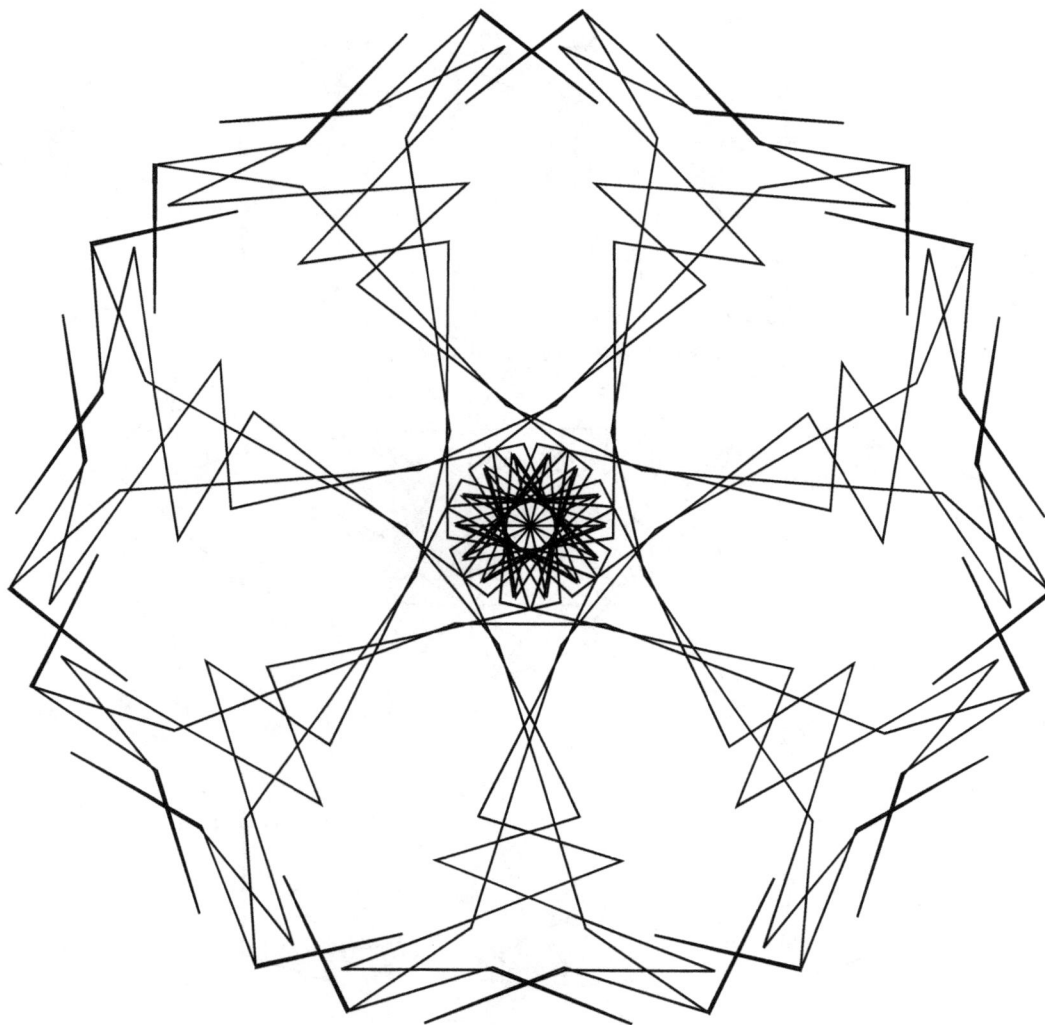

Figure 2.17 AAC [378: 5 377 231 21]

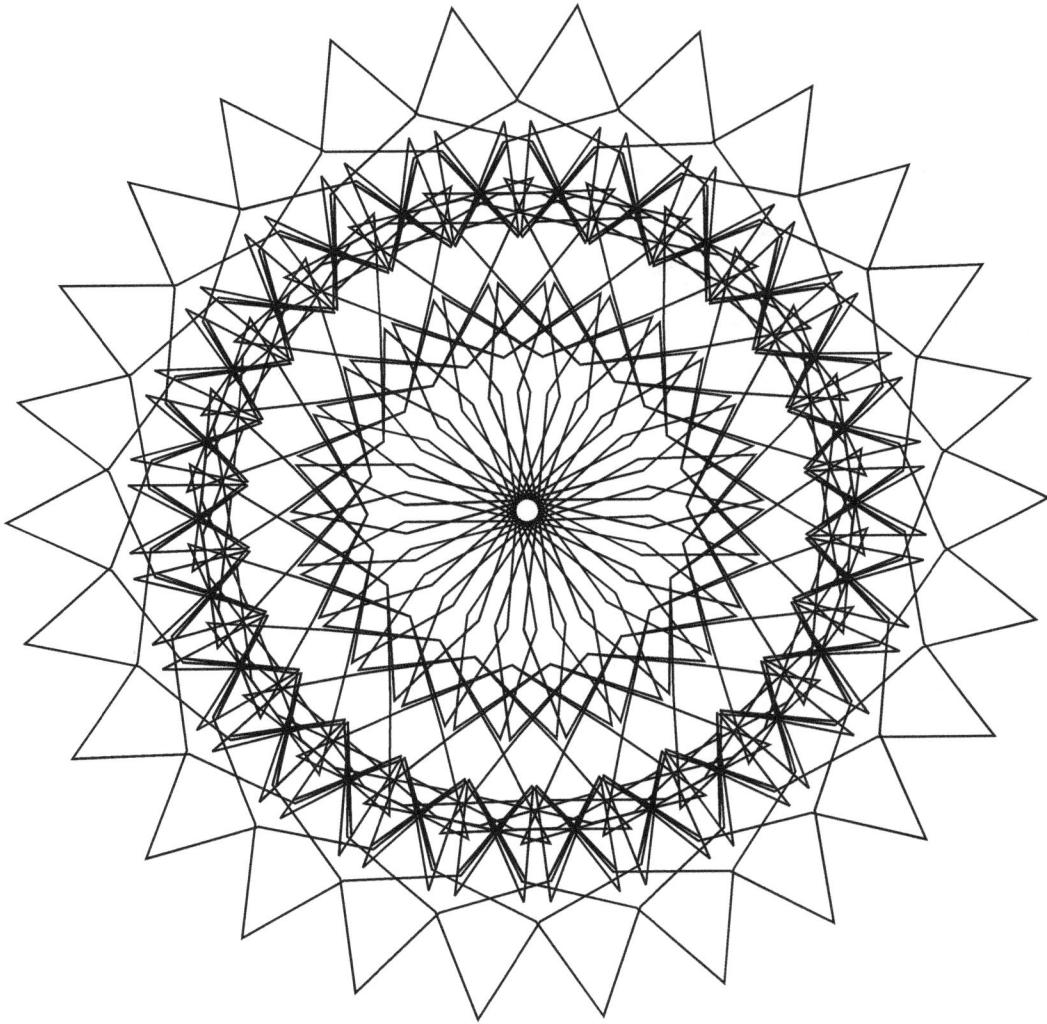

Figure 2.18   AAC [364: 0 28 104 130 182]

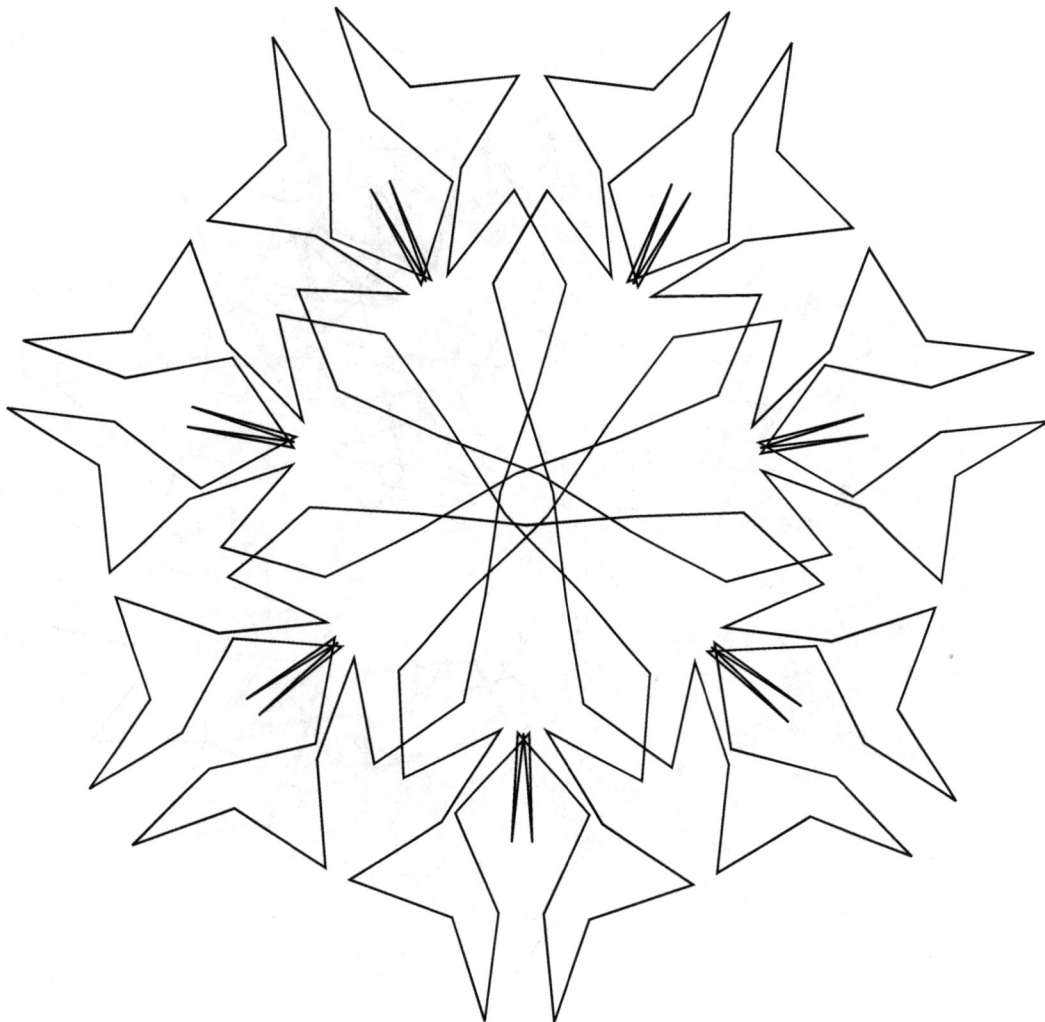

Figure 2.19 AAC [364: 7 129 182 28]

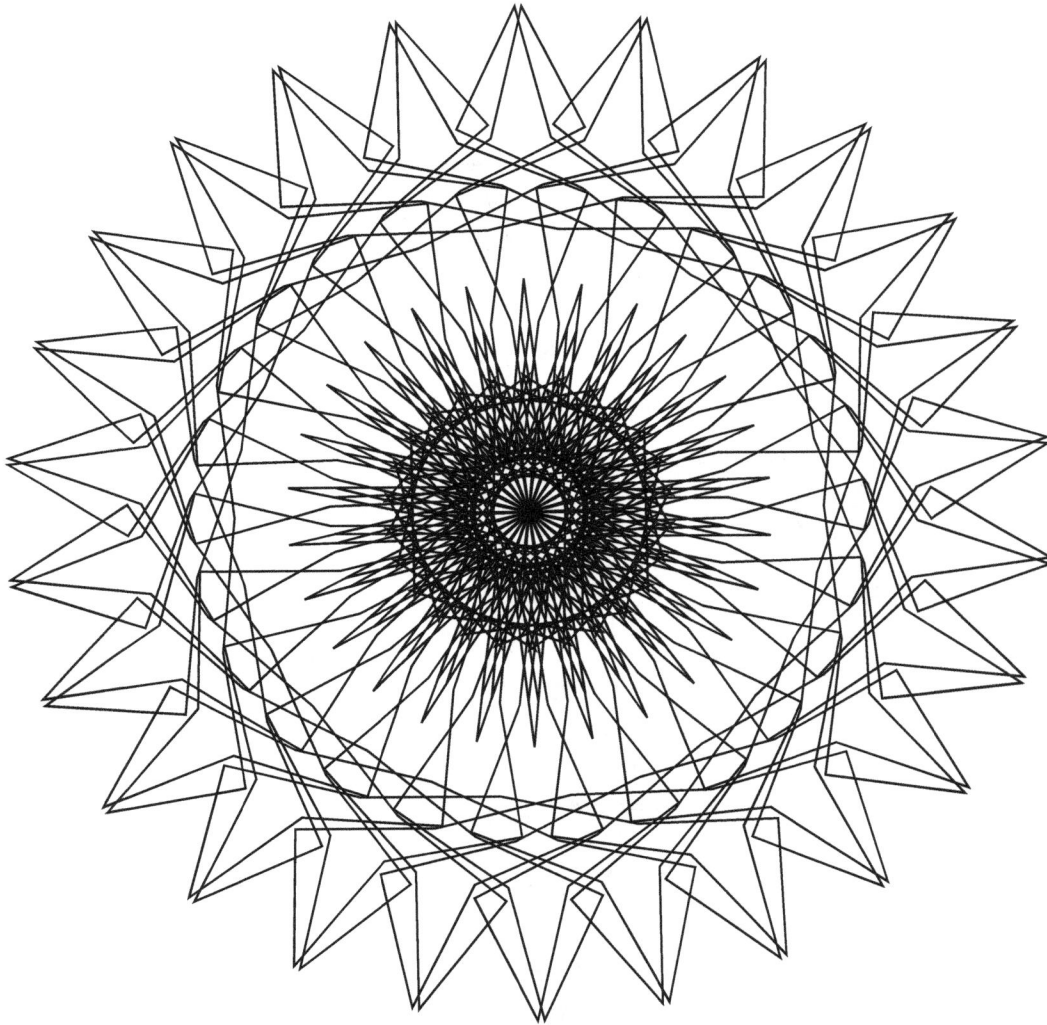

Figure 2.20 AAC [364: 0 350 286 312 182]

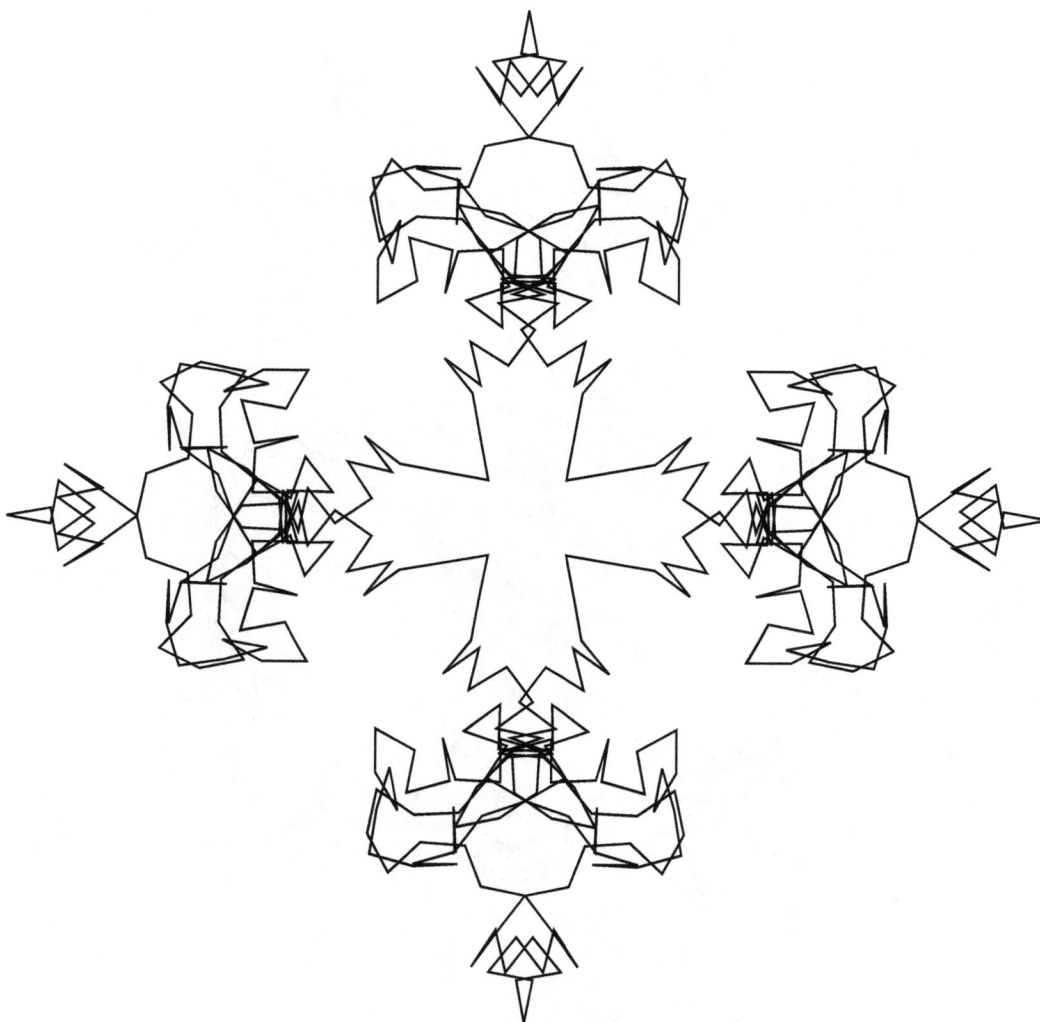

Figure 2.21   AAC [244: 58 139 205 94]

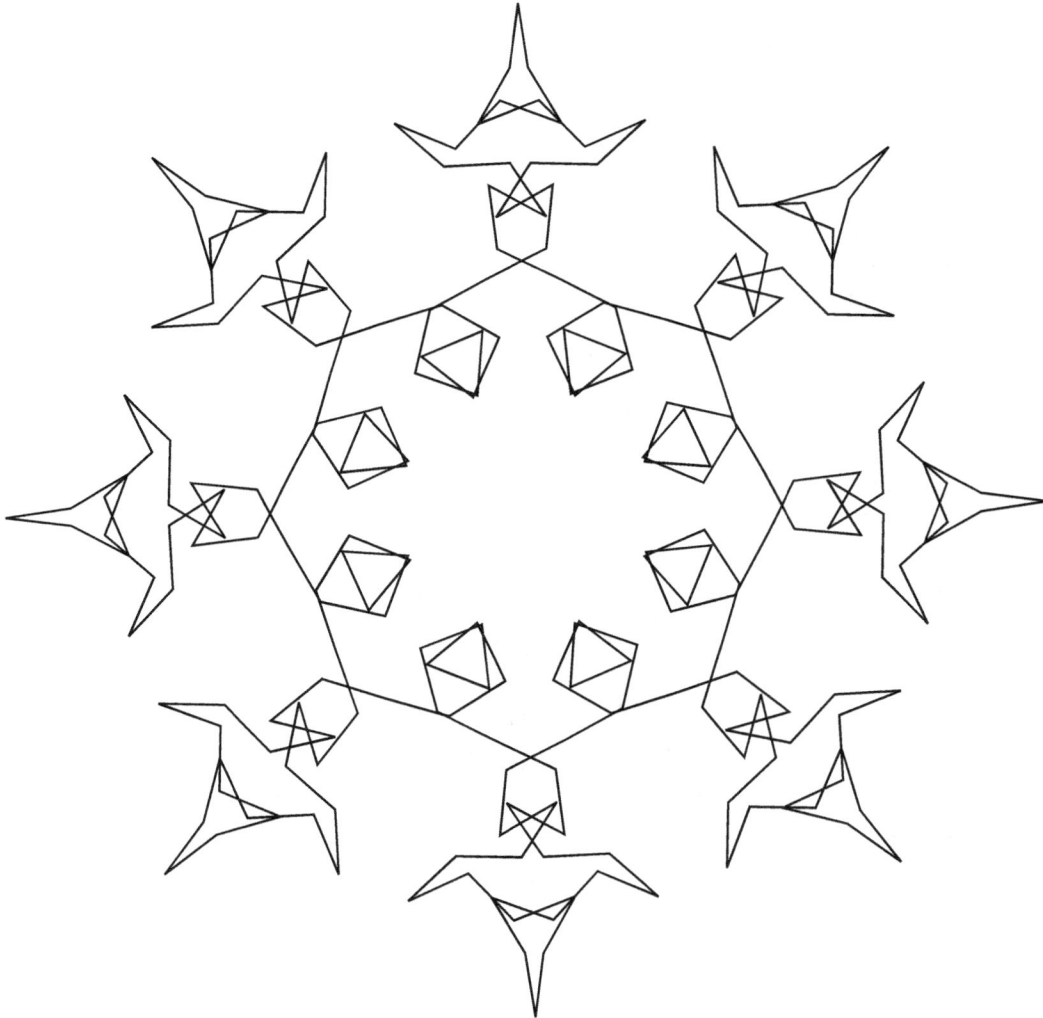

Figure 2.22   AAC [248: 10 33 216 240]

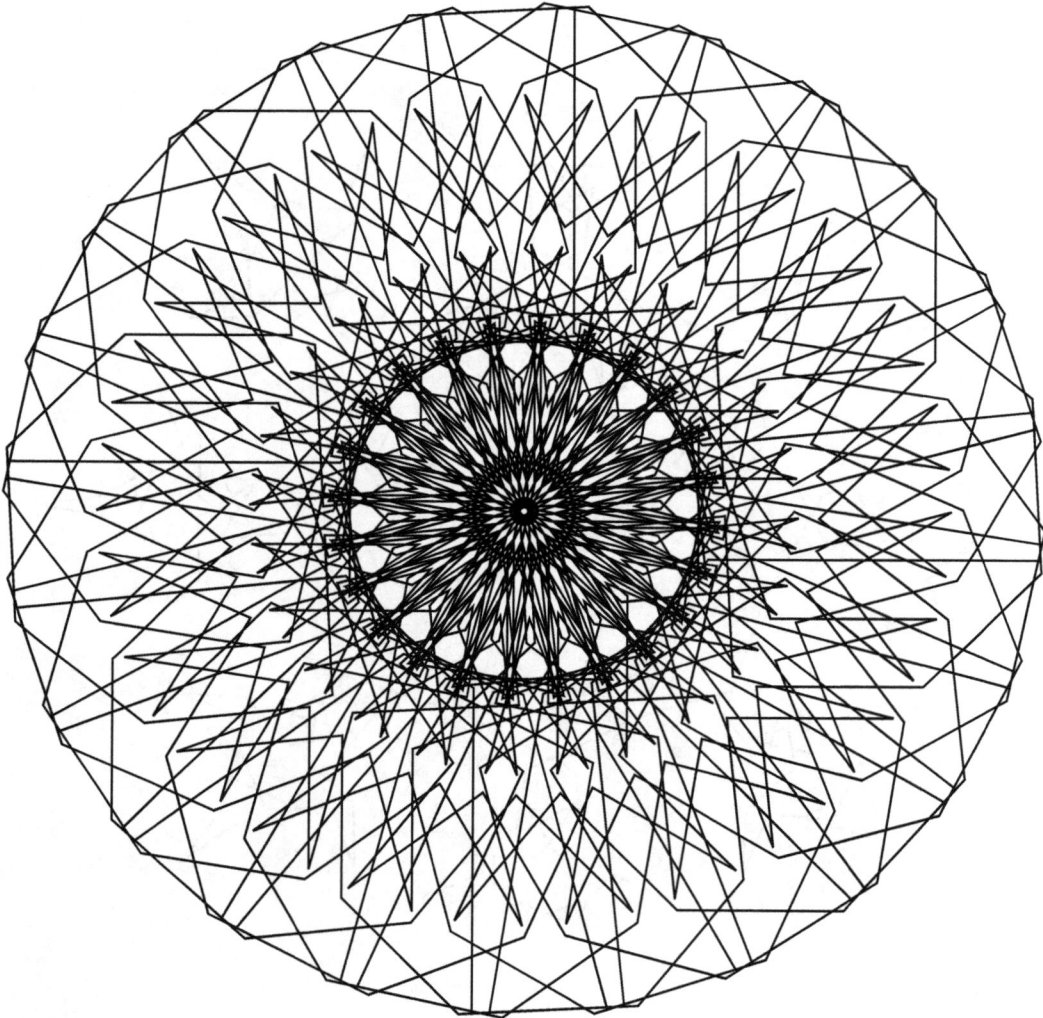

Figure 2.23 AAC [120: 0 15 1 14 60]

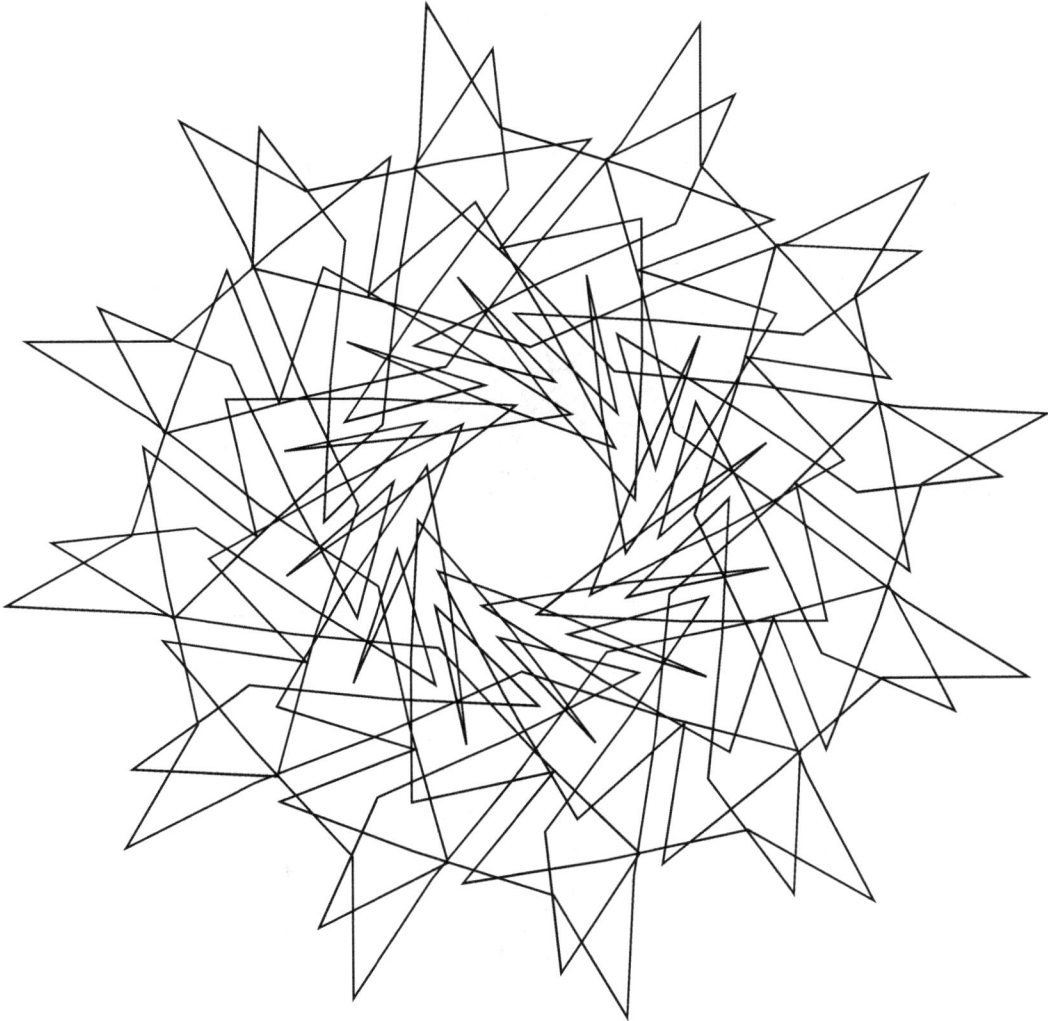

Figure 2.24   AAC [120: 0 34 39 54 48]

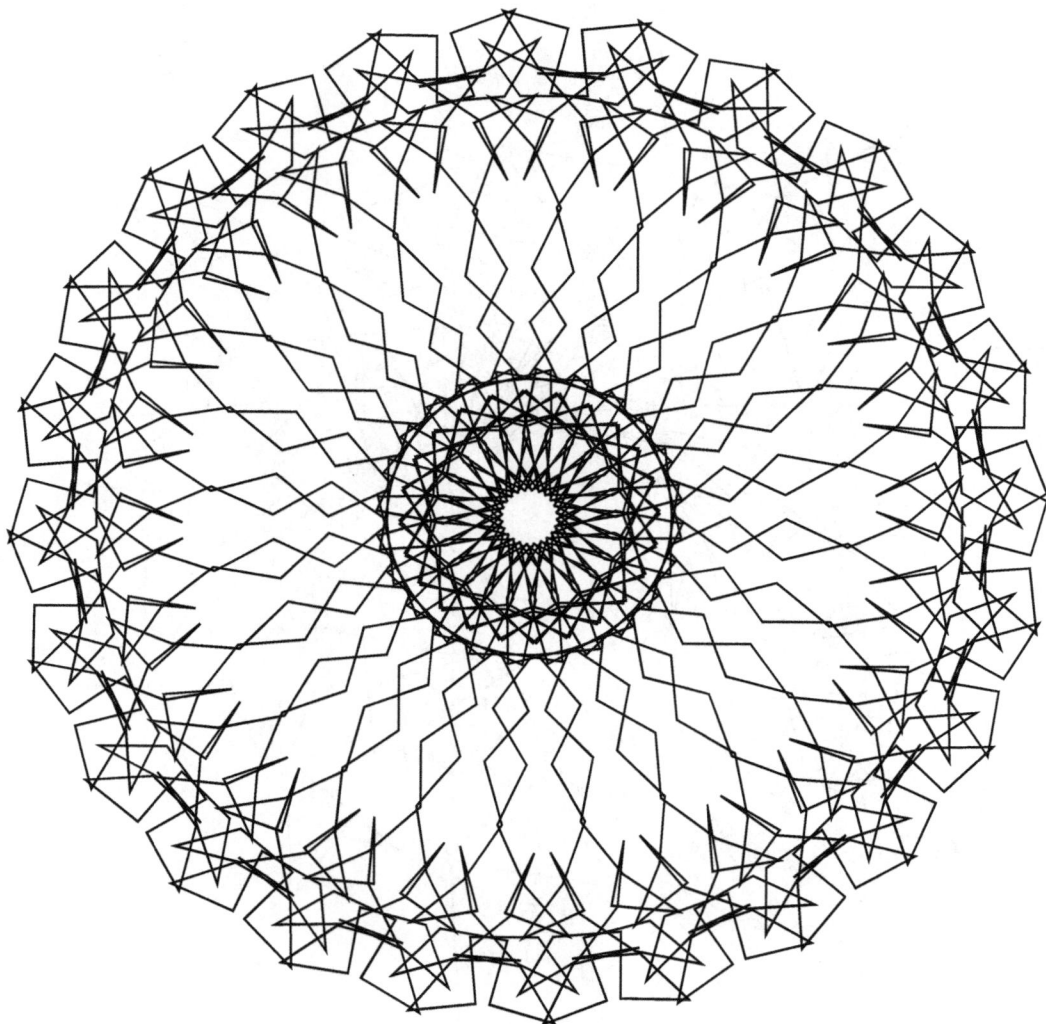

Figure 2.25   AAC [360: 0 175 216 136]

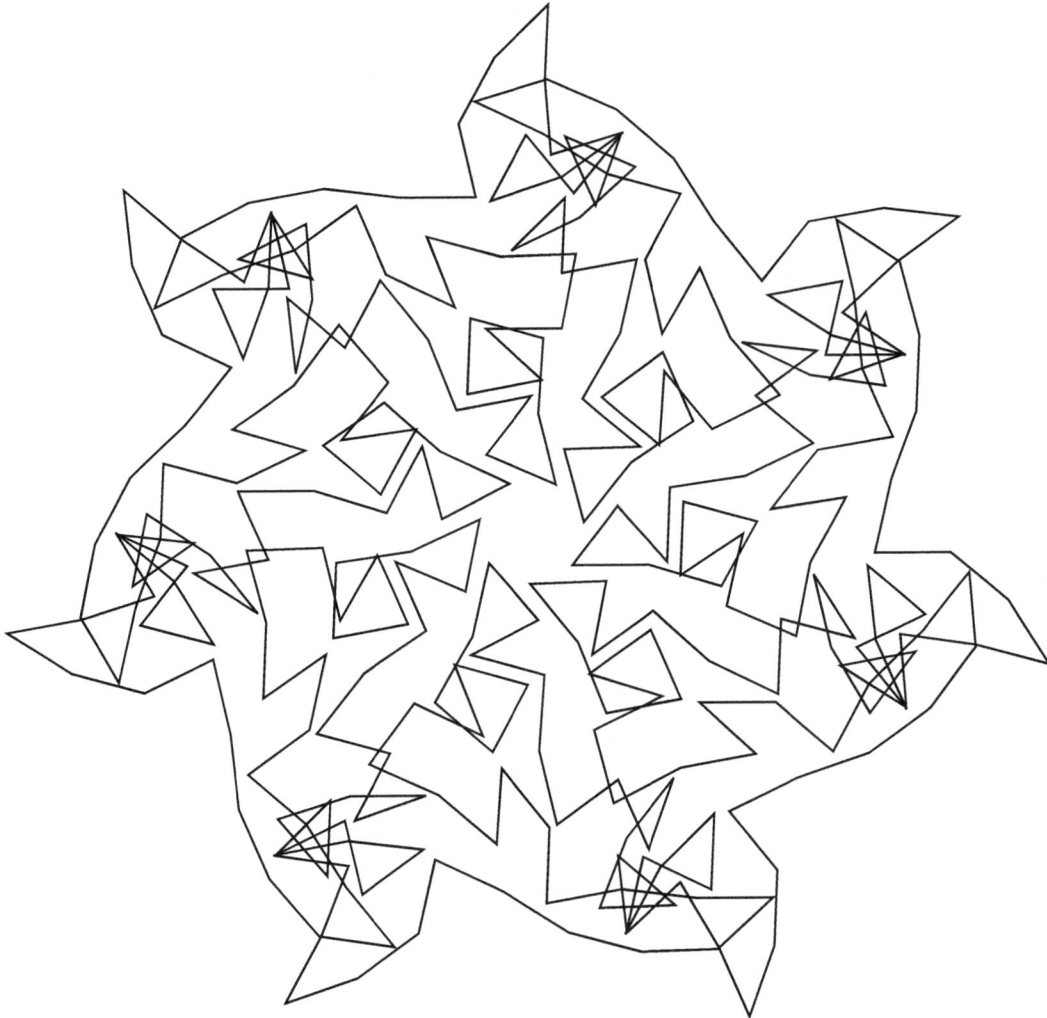

Figure 2.26   AAC [105: 0 30 63 49 14]

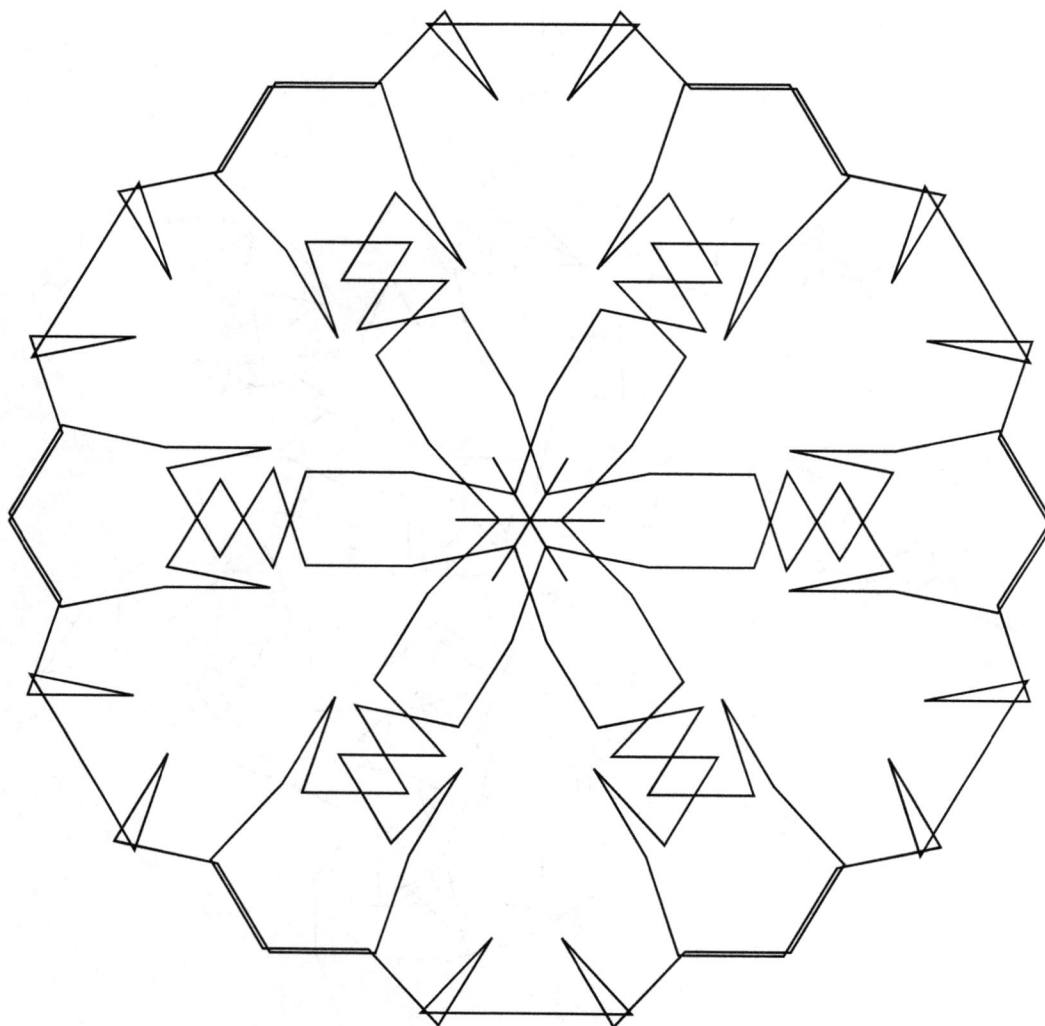

Figure 2.27   AAC [60: 0 28 54 44]

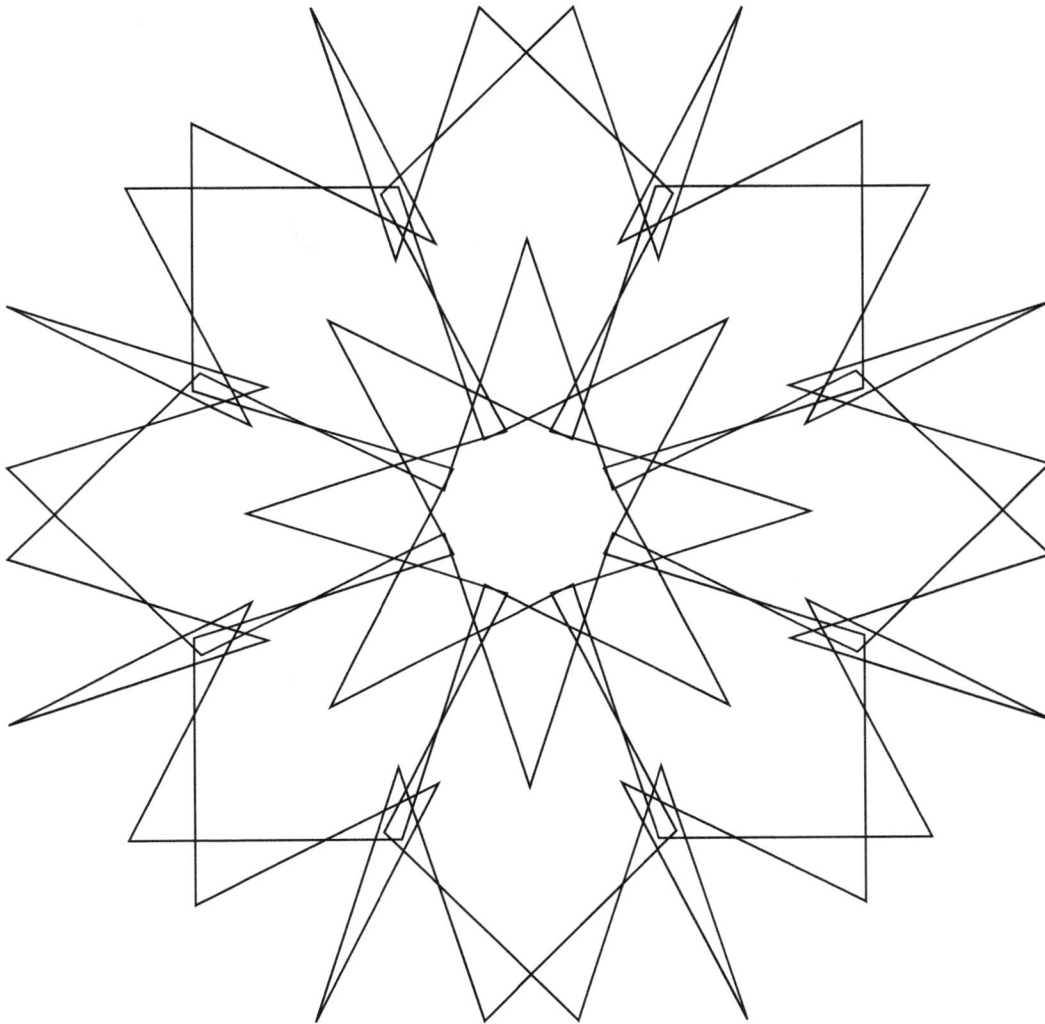

Figure 2.28   AAC [80: 0 26 22 52 40]

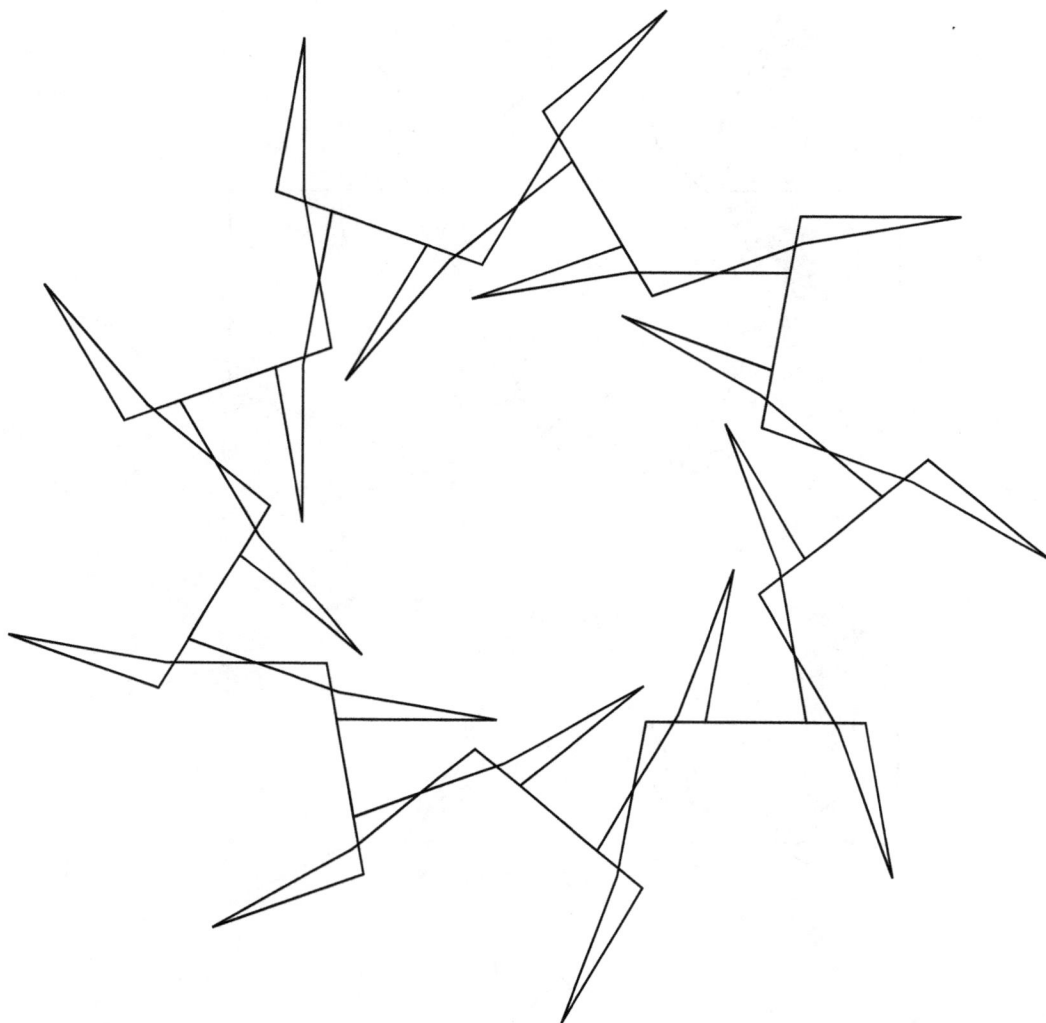

Figure 2.29   AAC [108: 0 57 54 27 54]

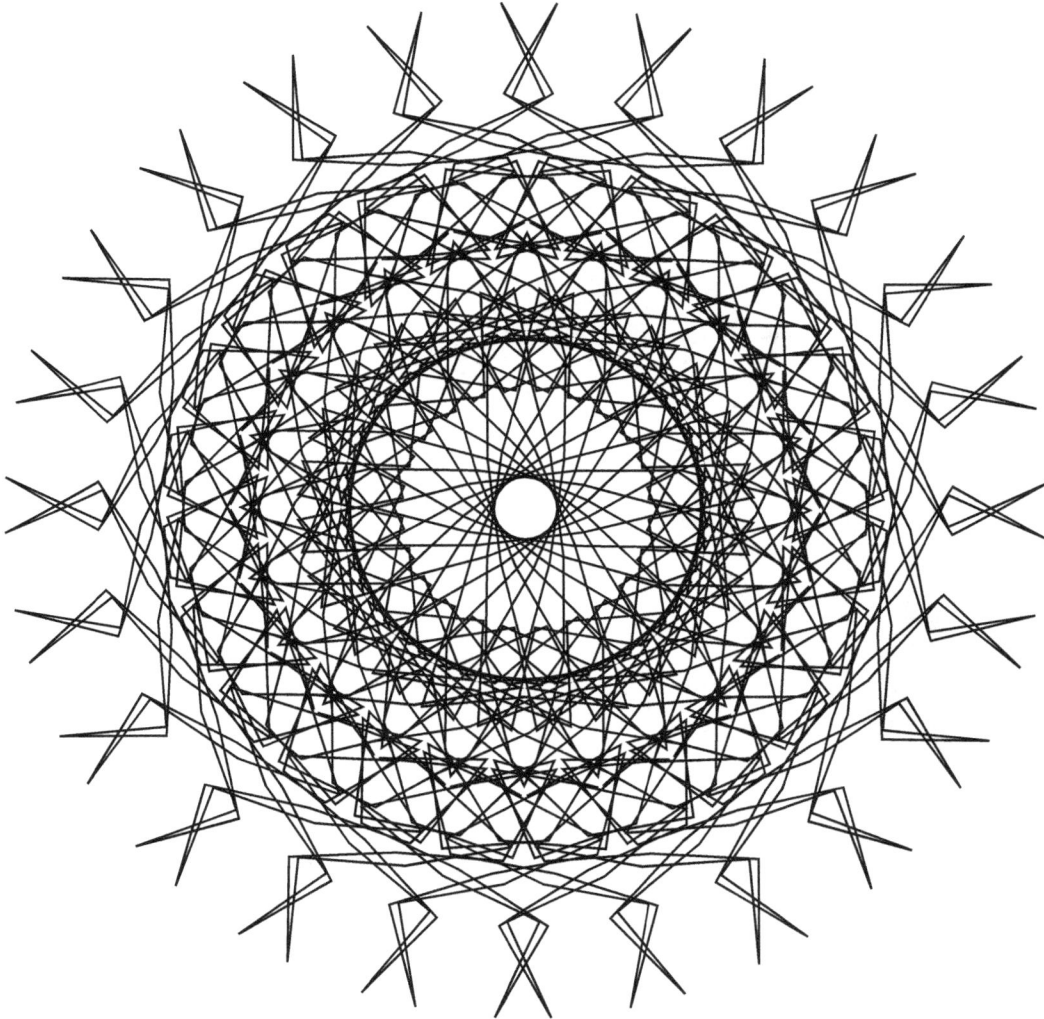

Figure 2.30   AAC [360: 0 7 0 8]

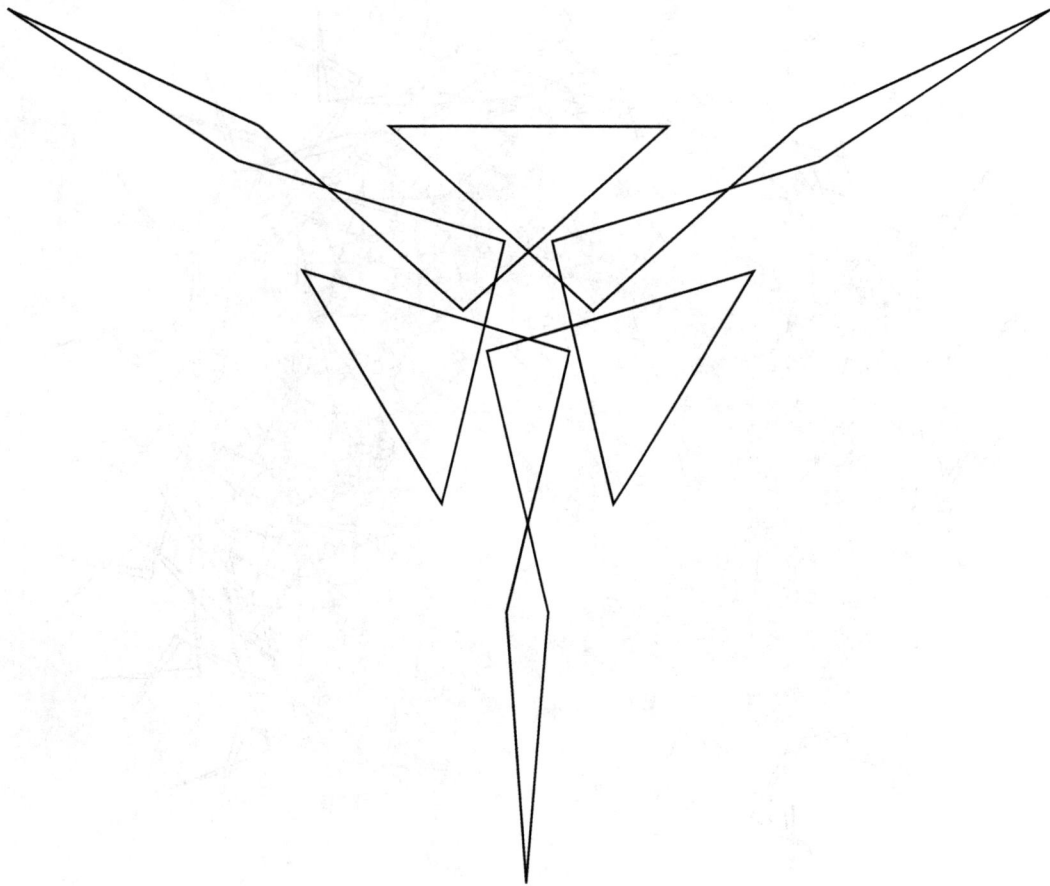

Figure 2.31   AAC [21: 1 13 0 3]

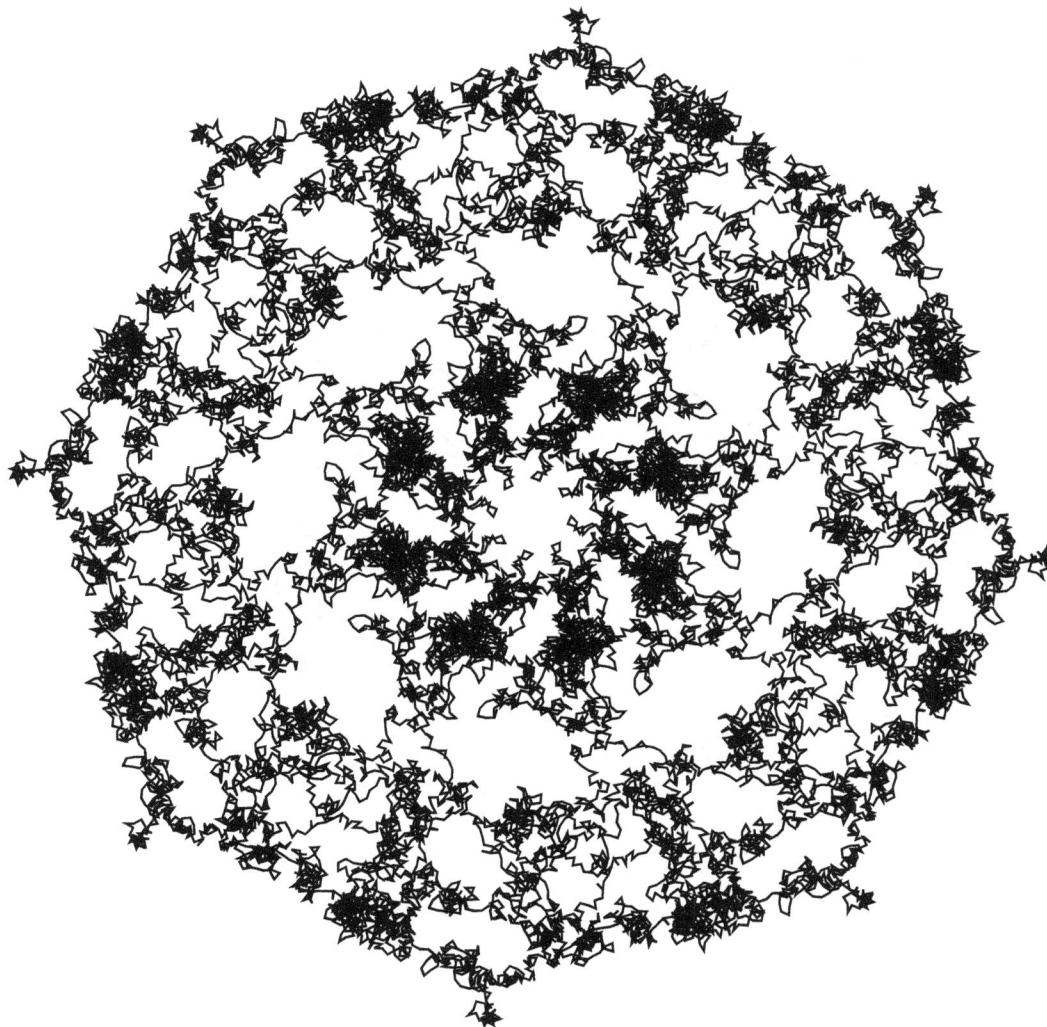

Figure 2.32 AAC [360: 0 17 100 109 103 246 218 176 116 48]

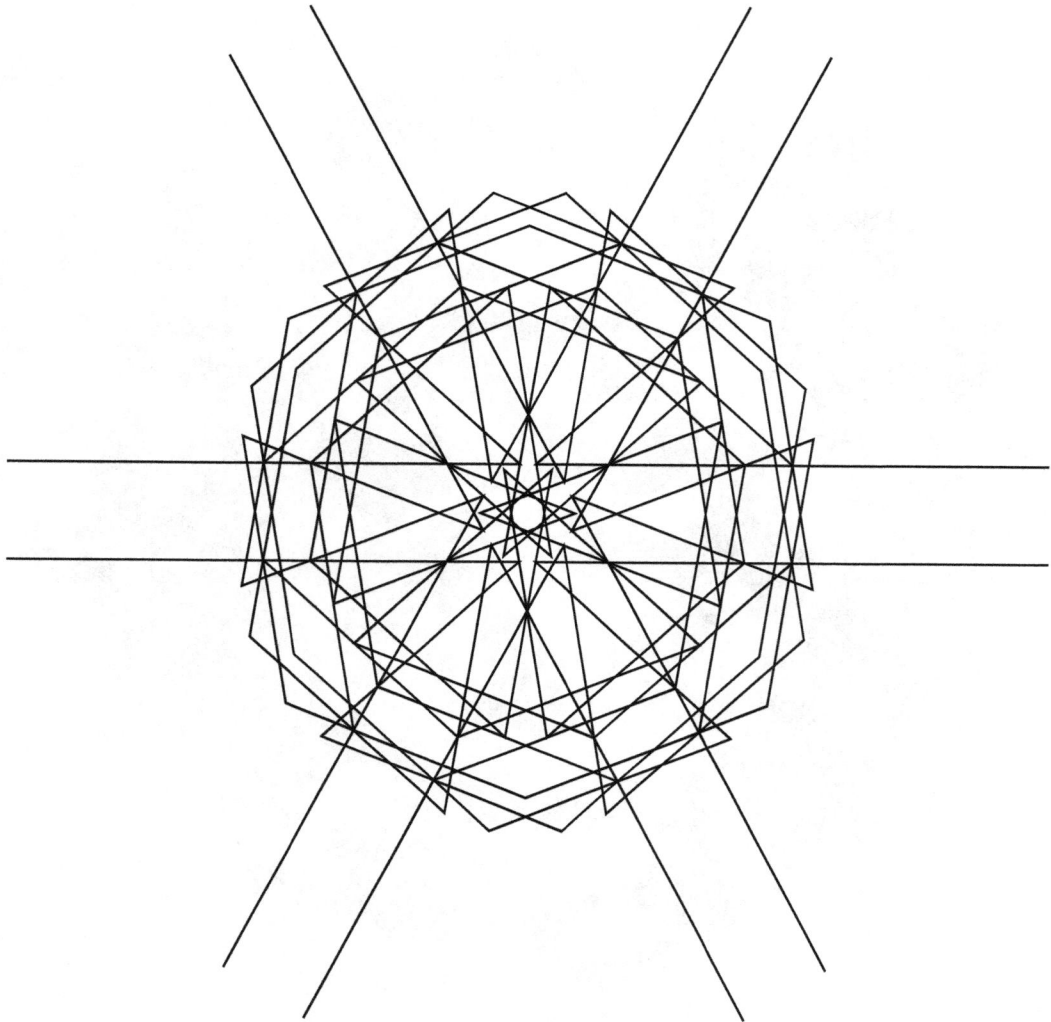

Figure 2.33   AAC [36: 2 12 2 32]

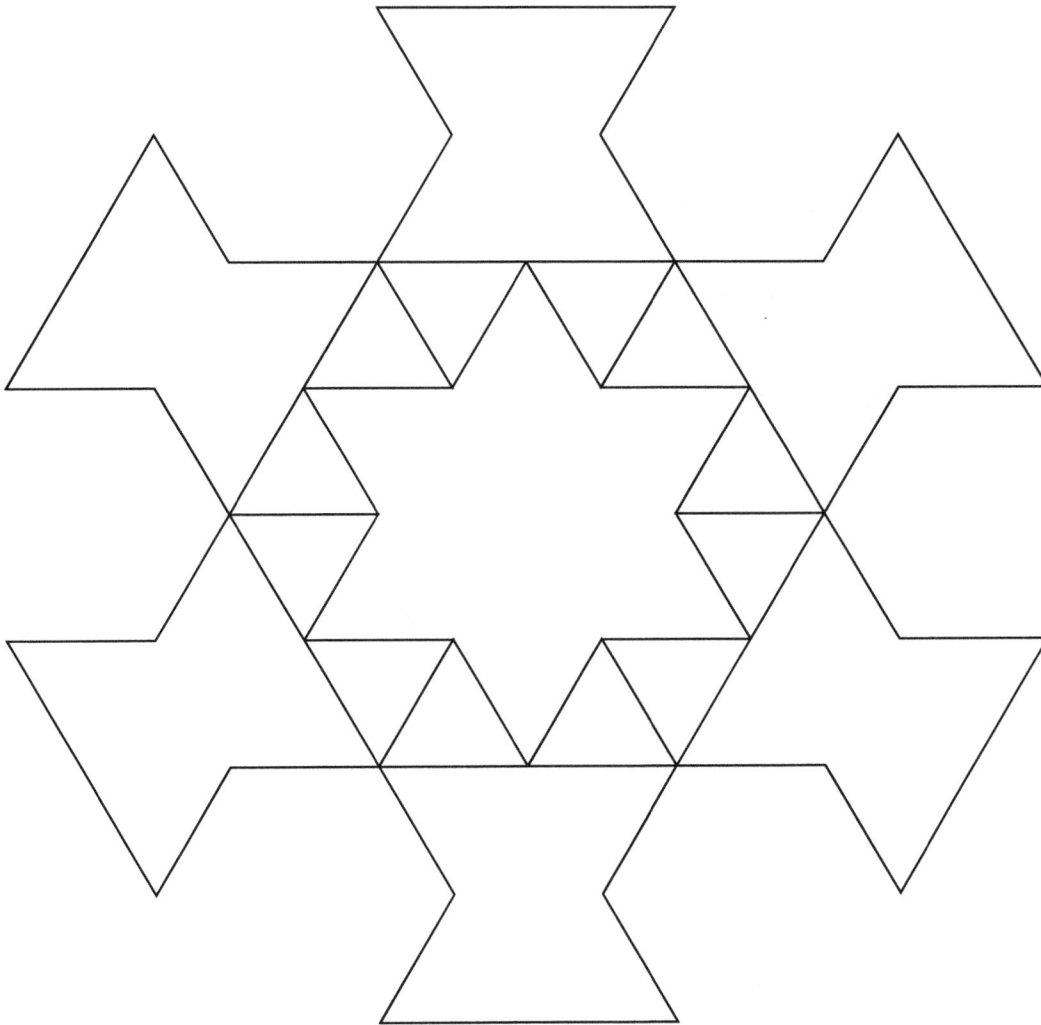

Figure 2.34 AAC [12: 0 4 0 2 6]

Figure 2.35   AAC [360: 3 157 120 68]

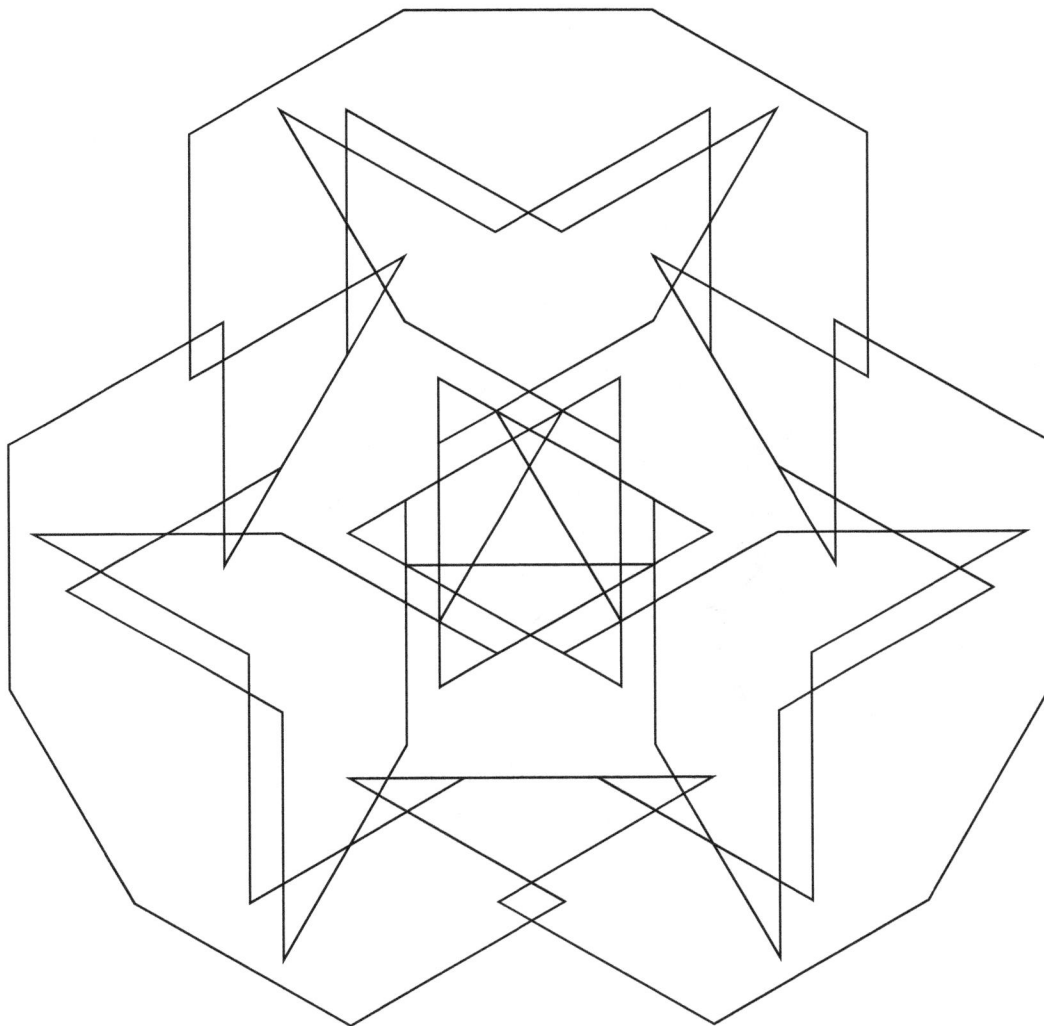

Figure 2.36   AAC [12: 3 10 10 1]

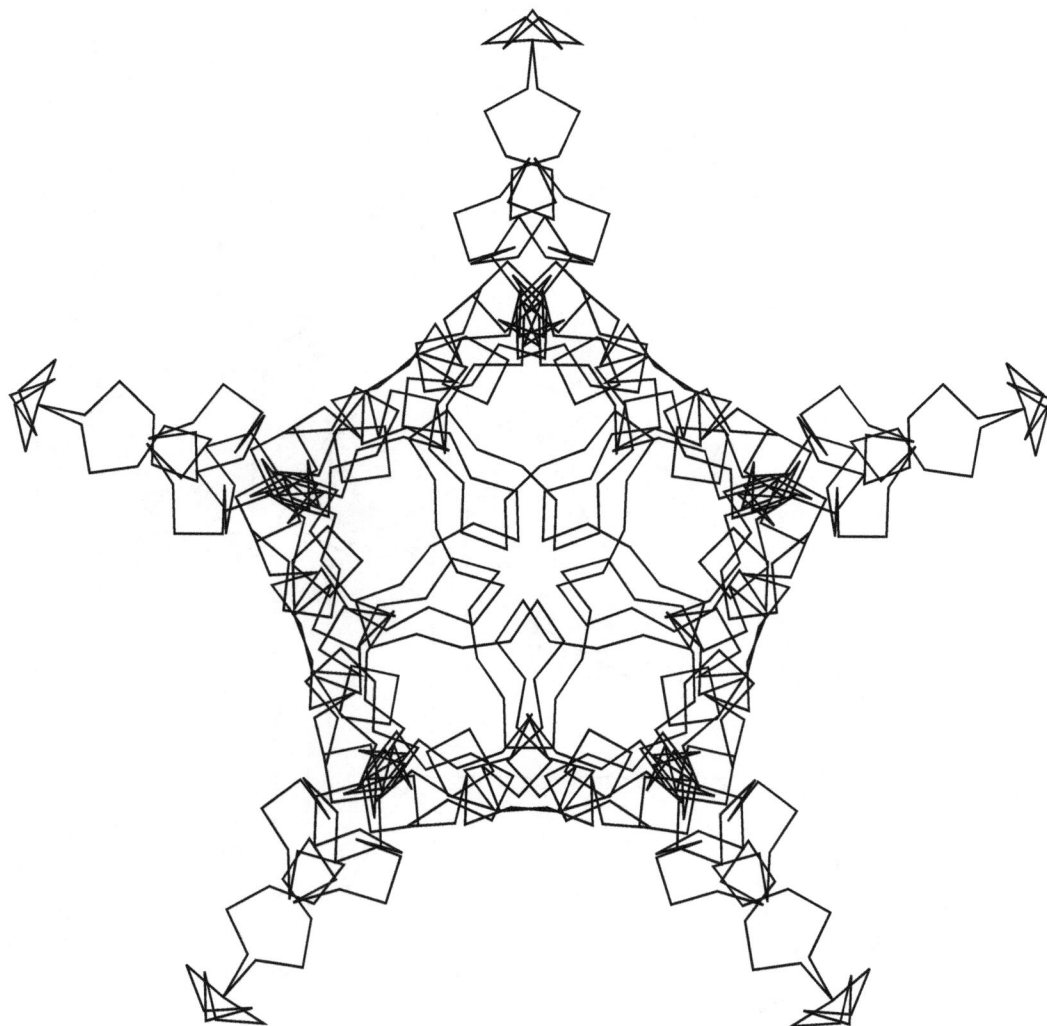

Figure 2.37   AAC [360: 64 88 0 216 192 296 280 280]

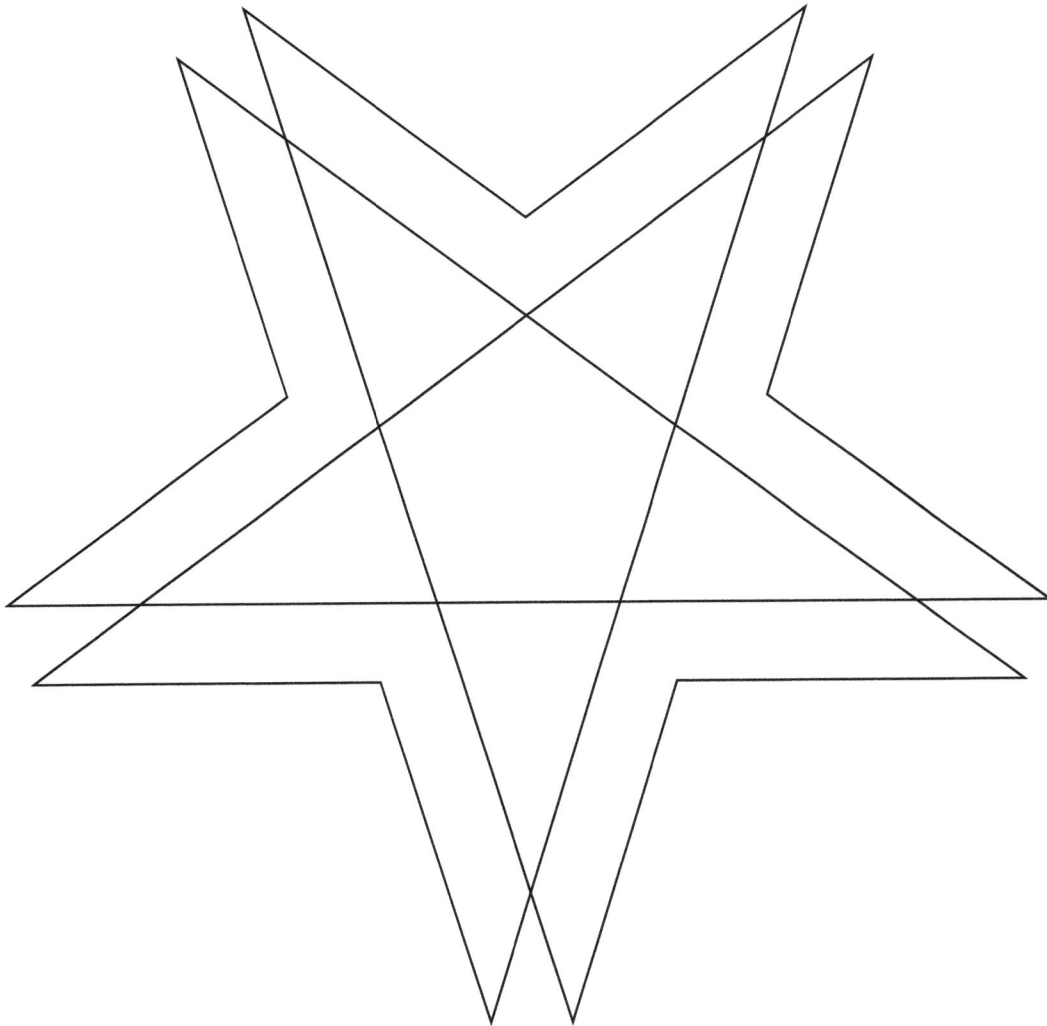

Figure 2.38   AAC [5: 0 2 3 2 0 3]

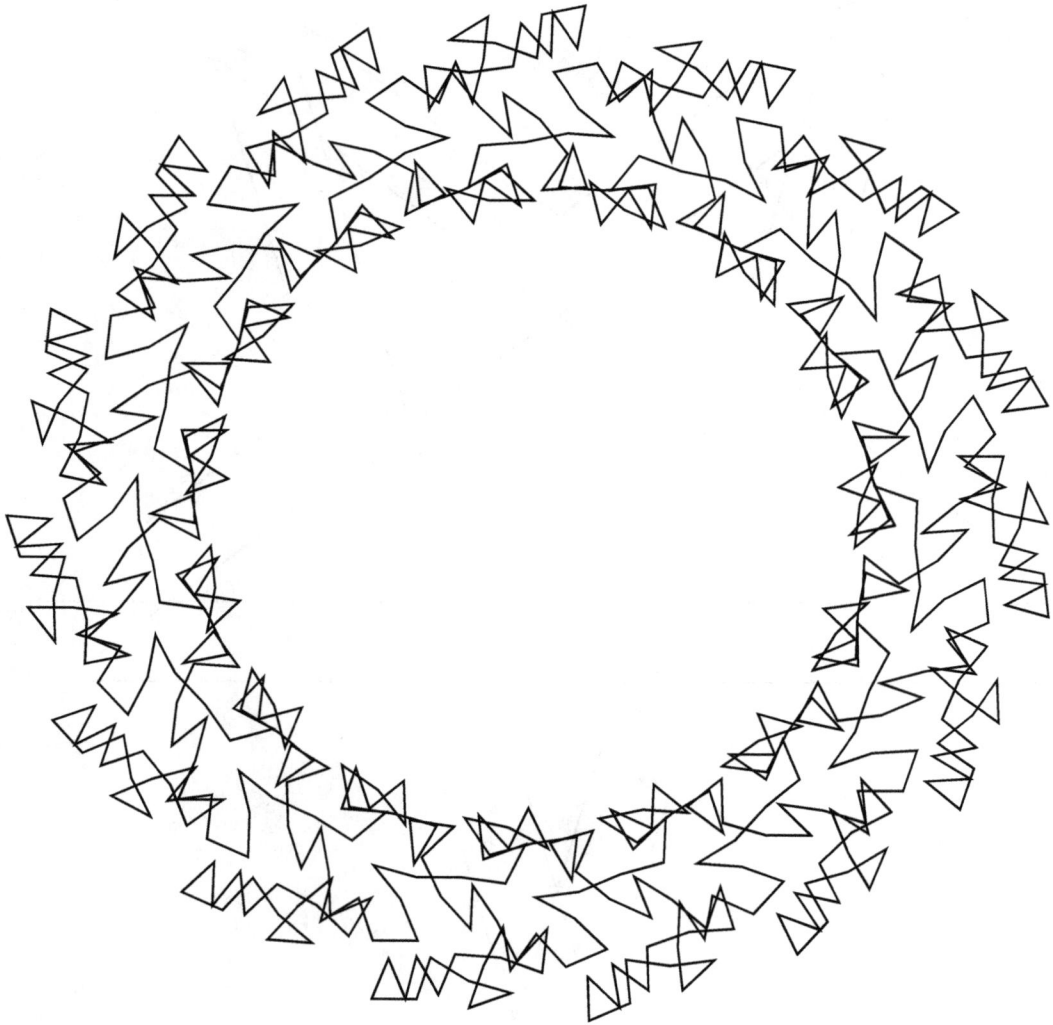

Figure 2.39   AAC [360: 0 216 40 80 216 312 240]

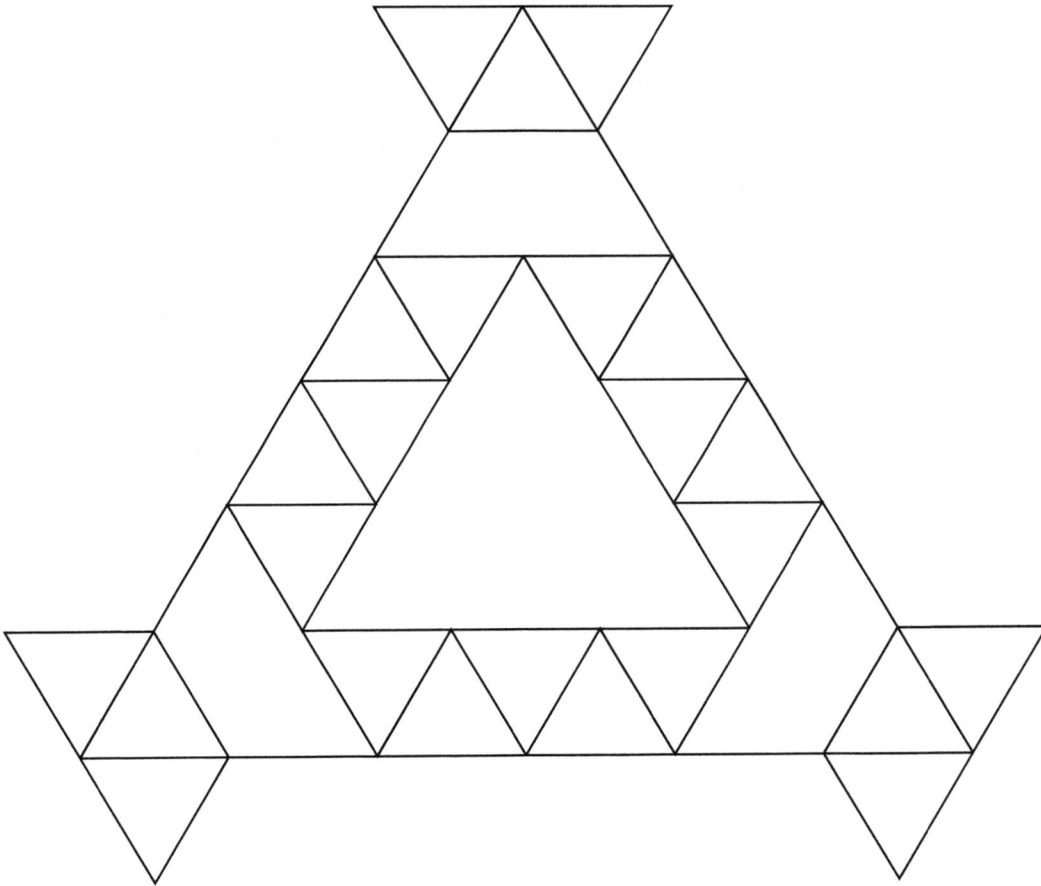

Figure 2.40 AAC [6: 0 1 1 2 3 3]

Figure 2.41 AAC [360: 0 199 117 314]

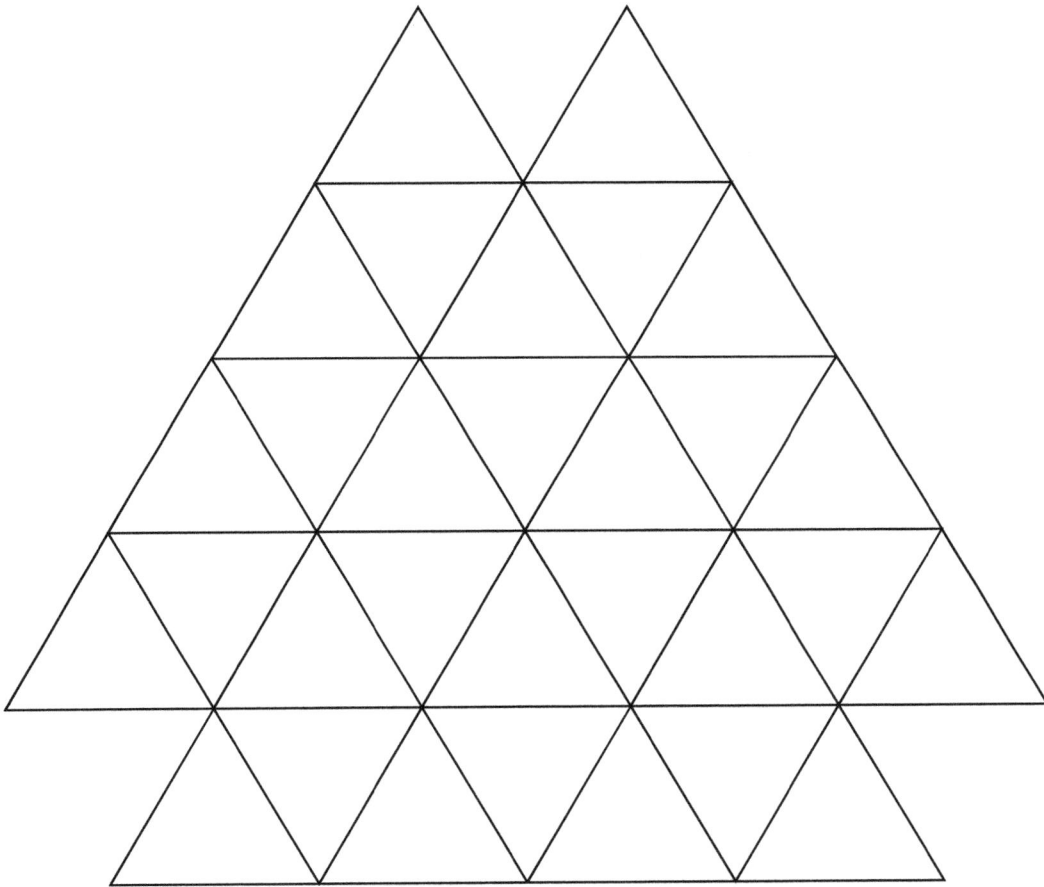

Figure 2.42 AAC [6: 0 1 1 4 3 3 0 3]

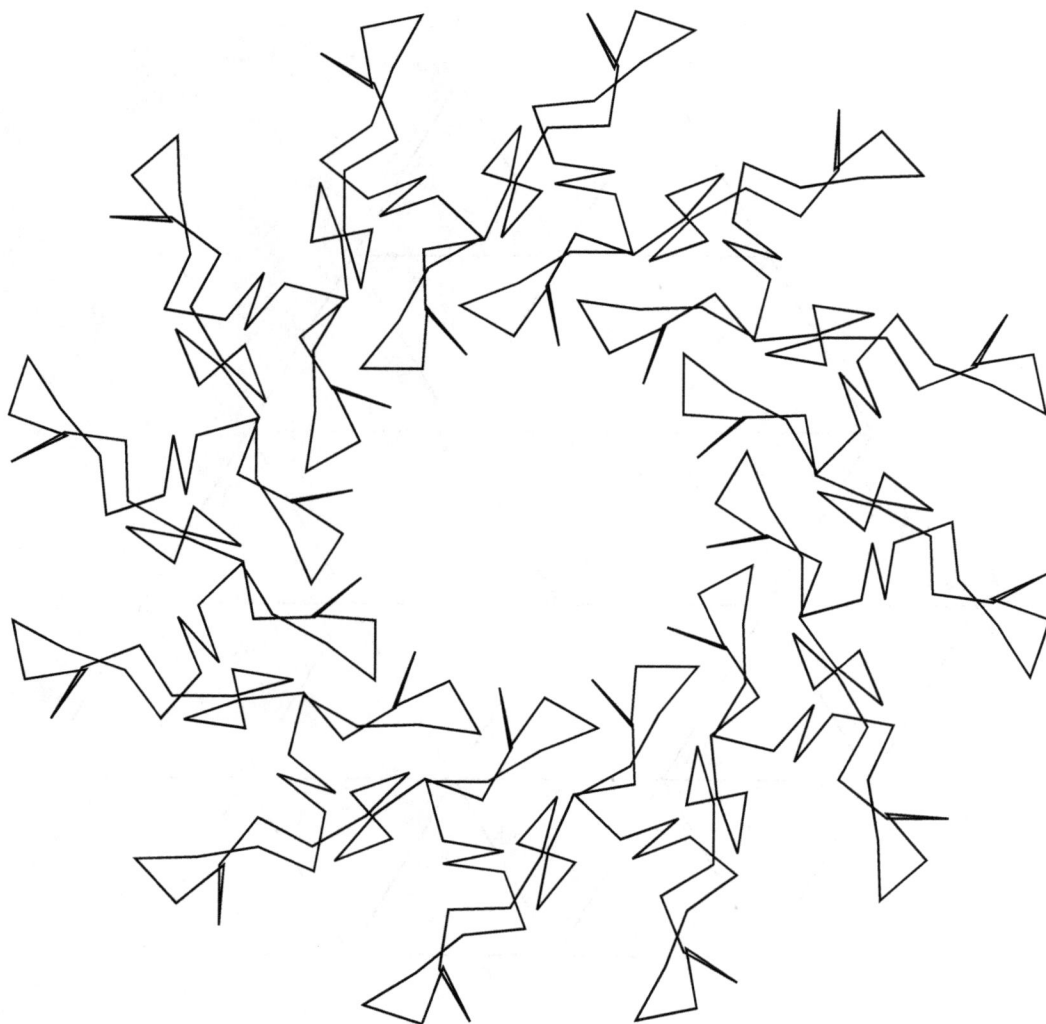

Figure 2.43 AAC [360: 0 302 138 180]

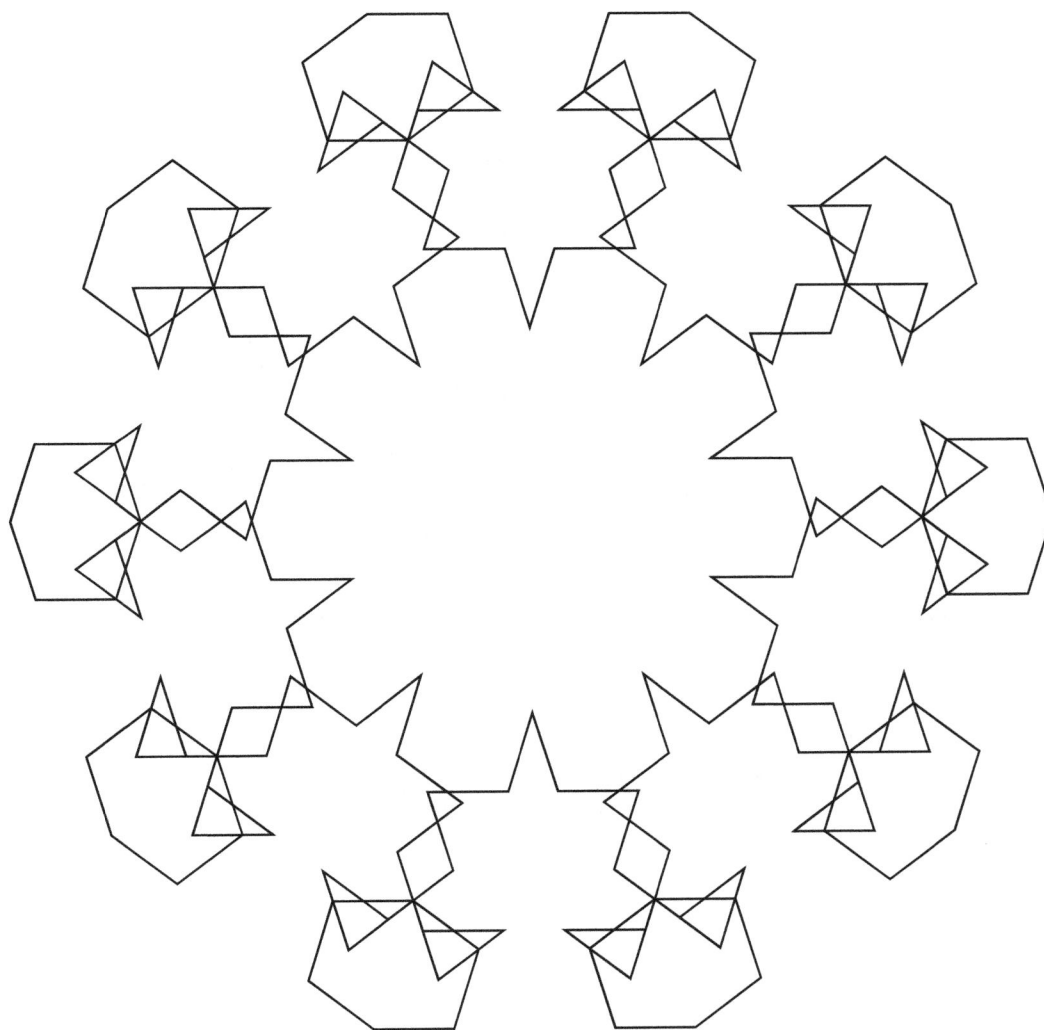

Figure 2.44   AAC [10: 0 8 6 8 3 6]

Figure 2.45   AAC [360: 0 131 156]

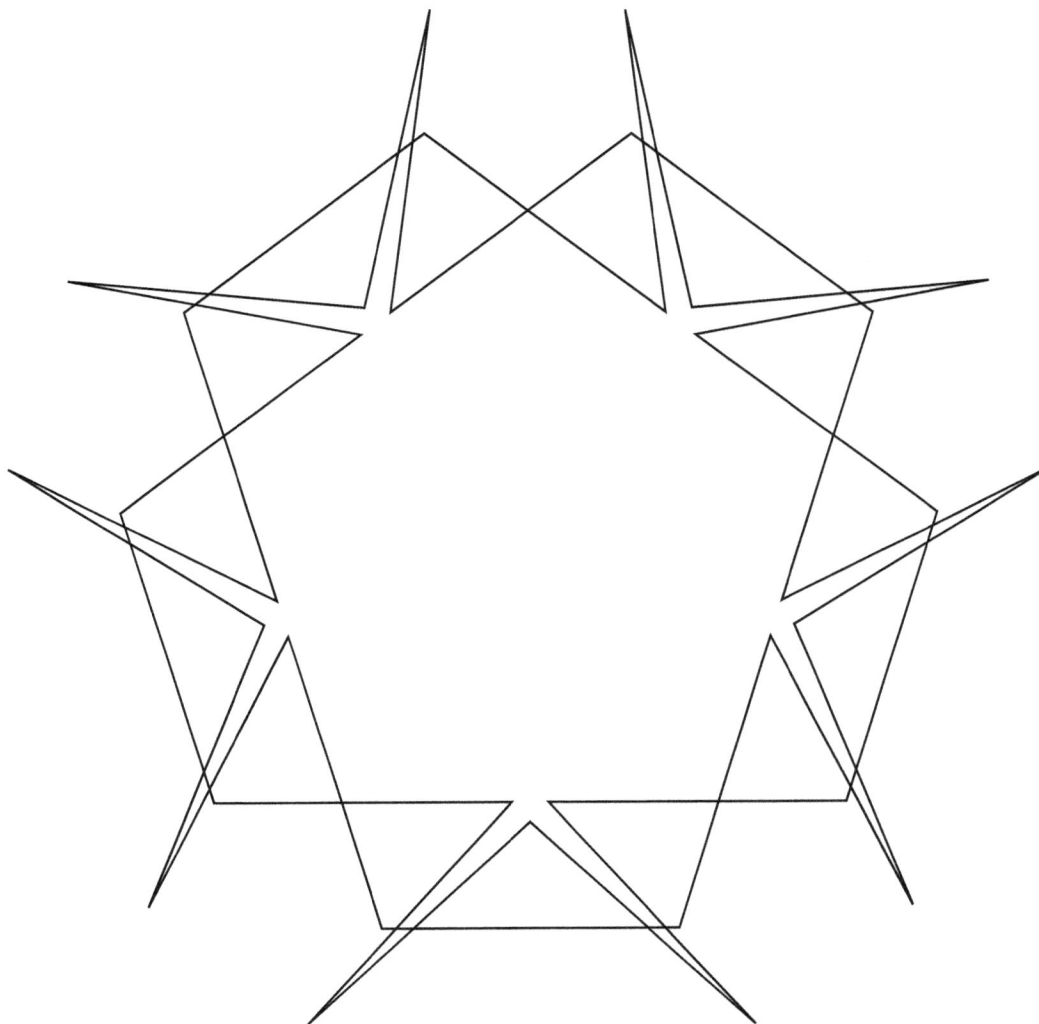

Figure 2.46 AAC [35: 1 17 10 15]

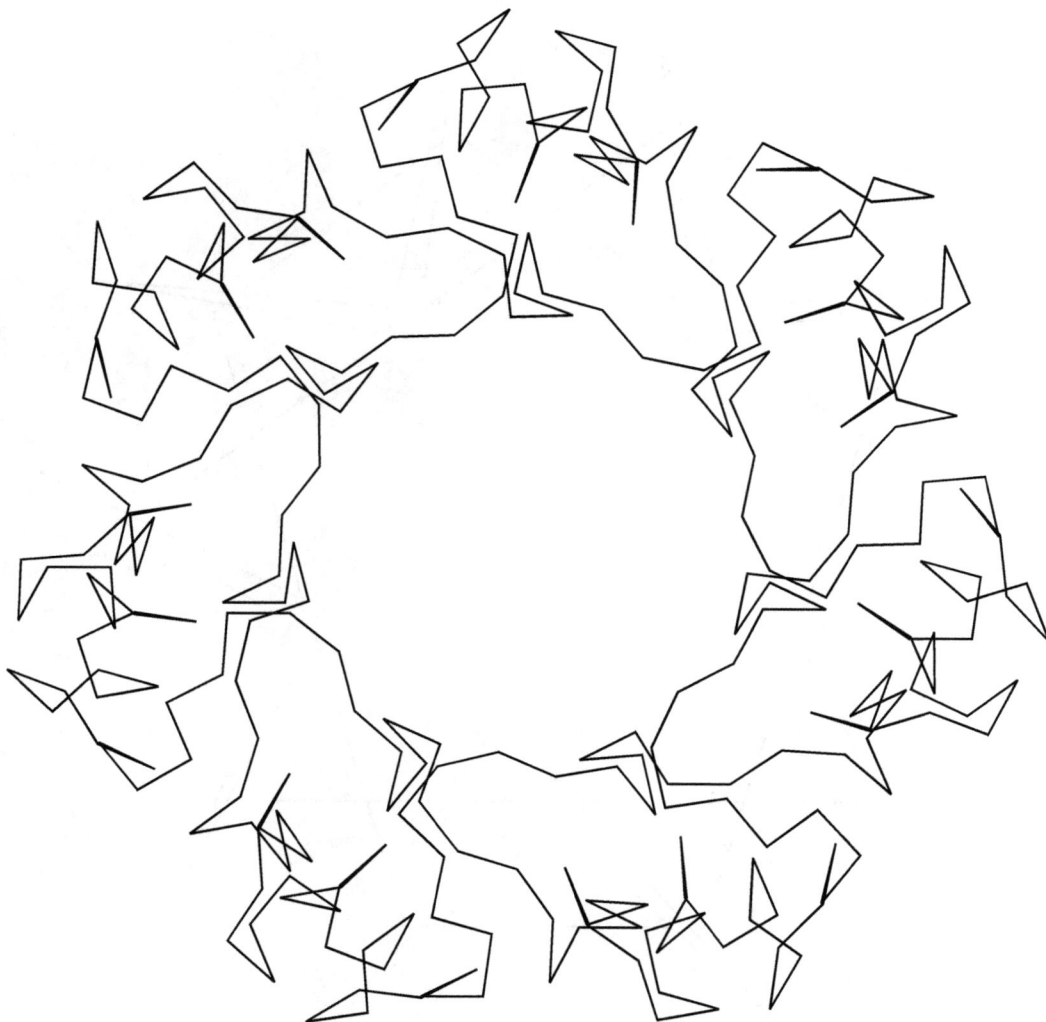

Figure 2.47   AAC [105: 0 24 0 56 35]

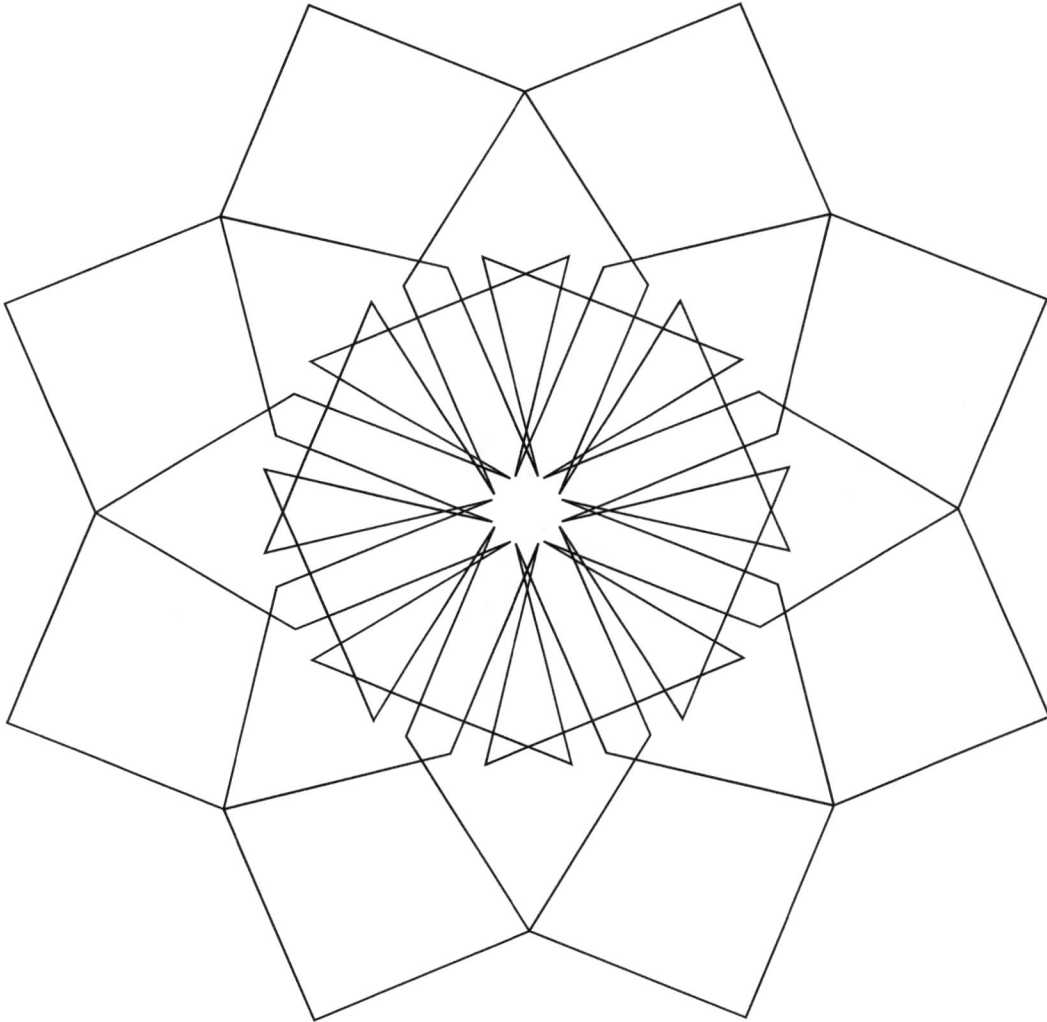

Figure 2.48   AAC [80: 3 52 18 44 40]

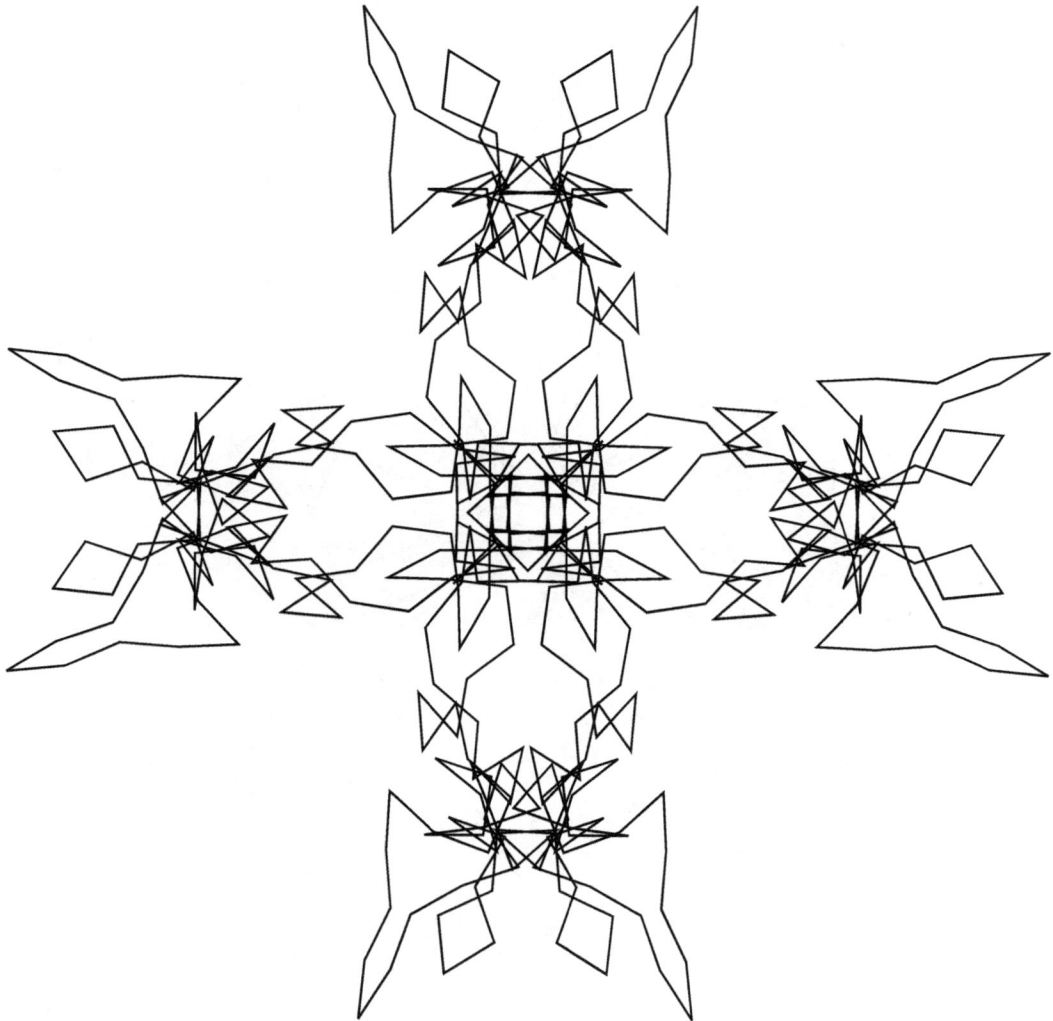

Figure 2.49   AAC [244: 58 87 65 186]

Figure 2.50   AAC [256: 0 37 252 40]

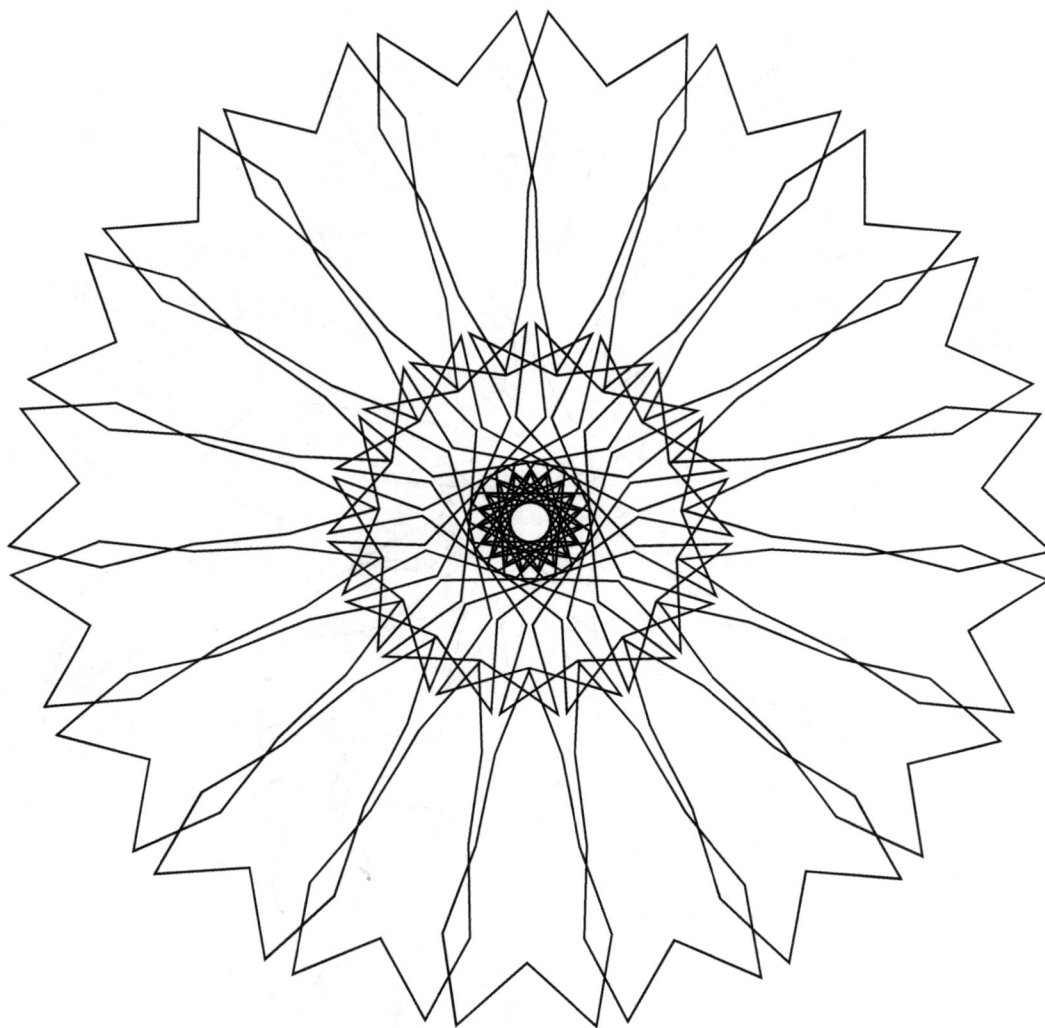

Figure 2.51   AAC [361: 4 332 38 304]

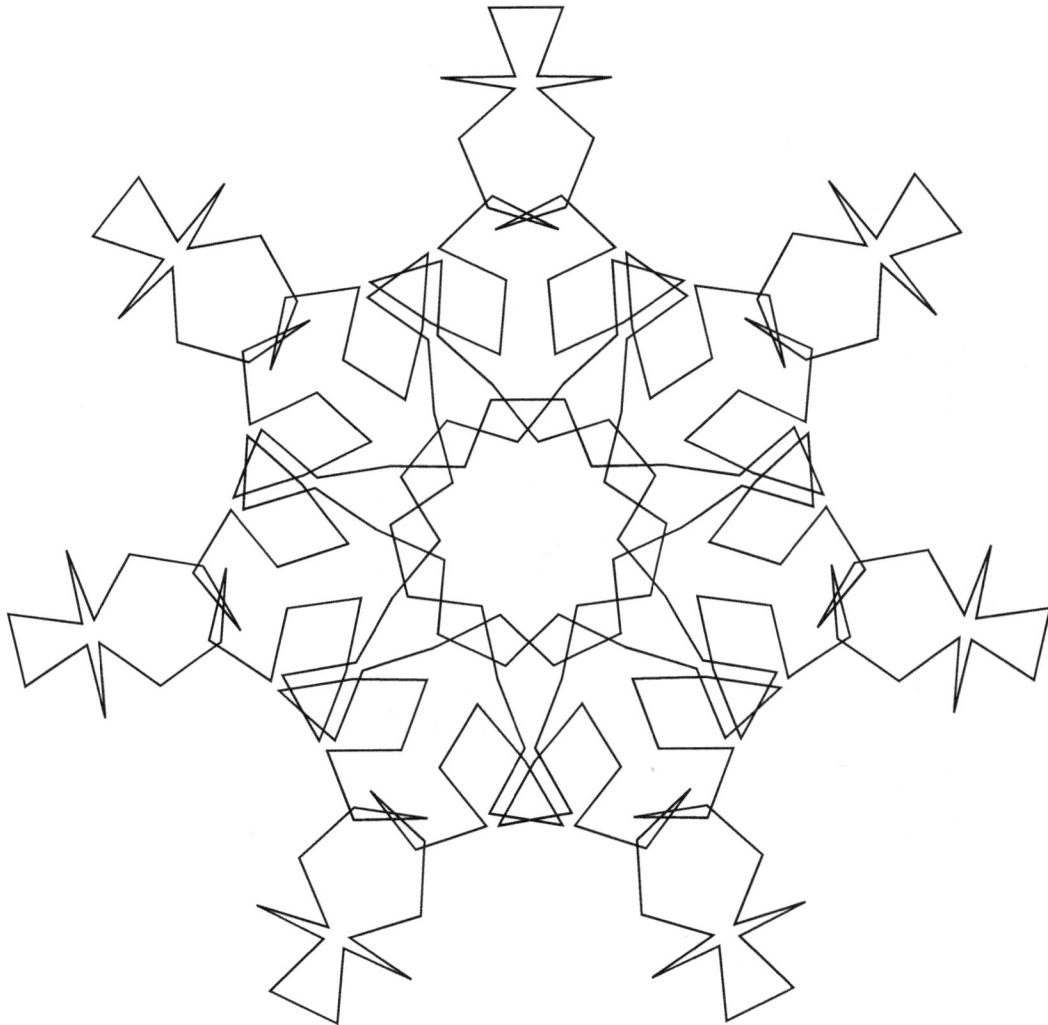

Figure 2.52   AAC [378: 9 368 84 147]

Figure 2.53   AAC [336: 0 31 219 2]

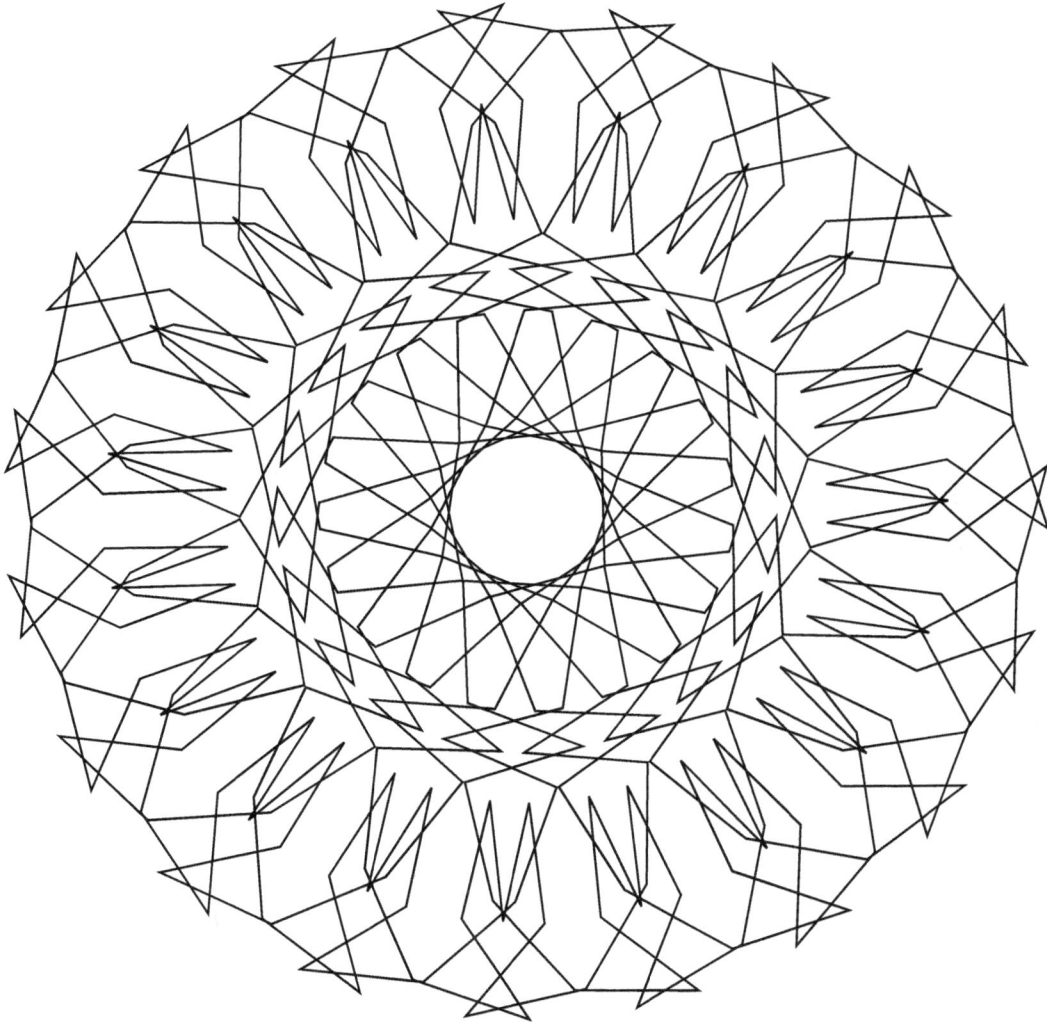

Figure 2.54   AAC [361: 0 218 95 285]

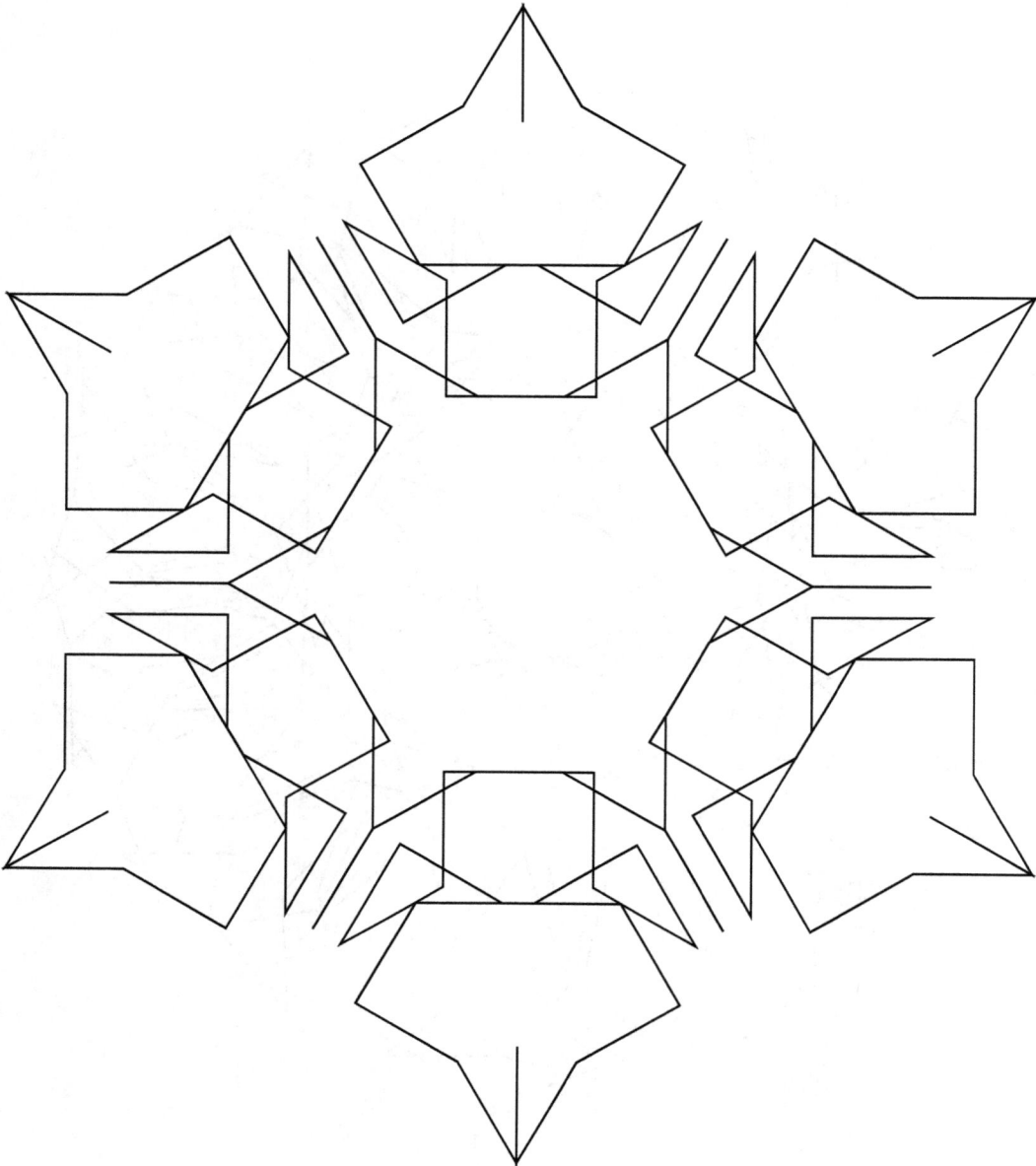

Figure 2.55 AAC [12: 0 1 8 5 9 6]

Figure 2.56   AAC [360: 0 0 0 0 0 9]

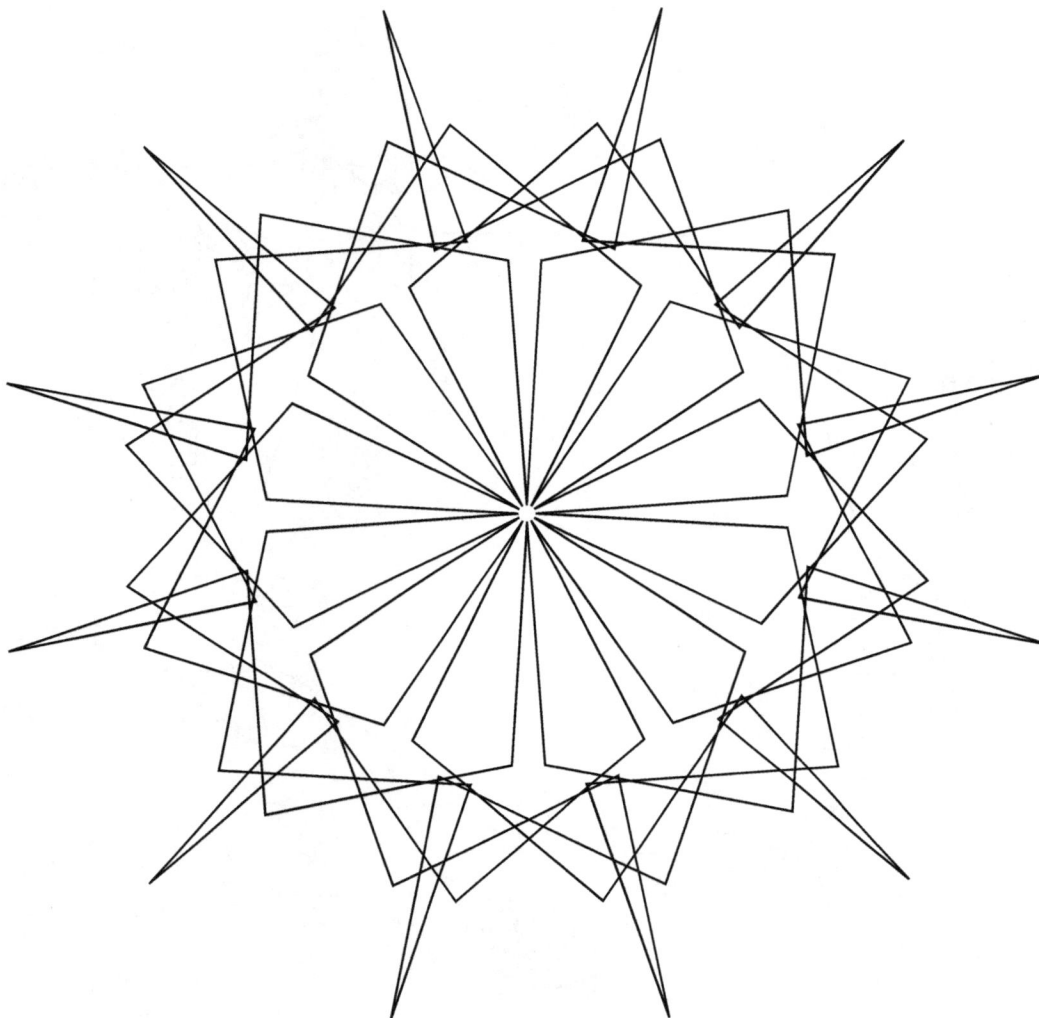

Figure 2.57   AAC [240: 8 175 135 150]

Figure 2.58 AAC [96: 0 1 56]

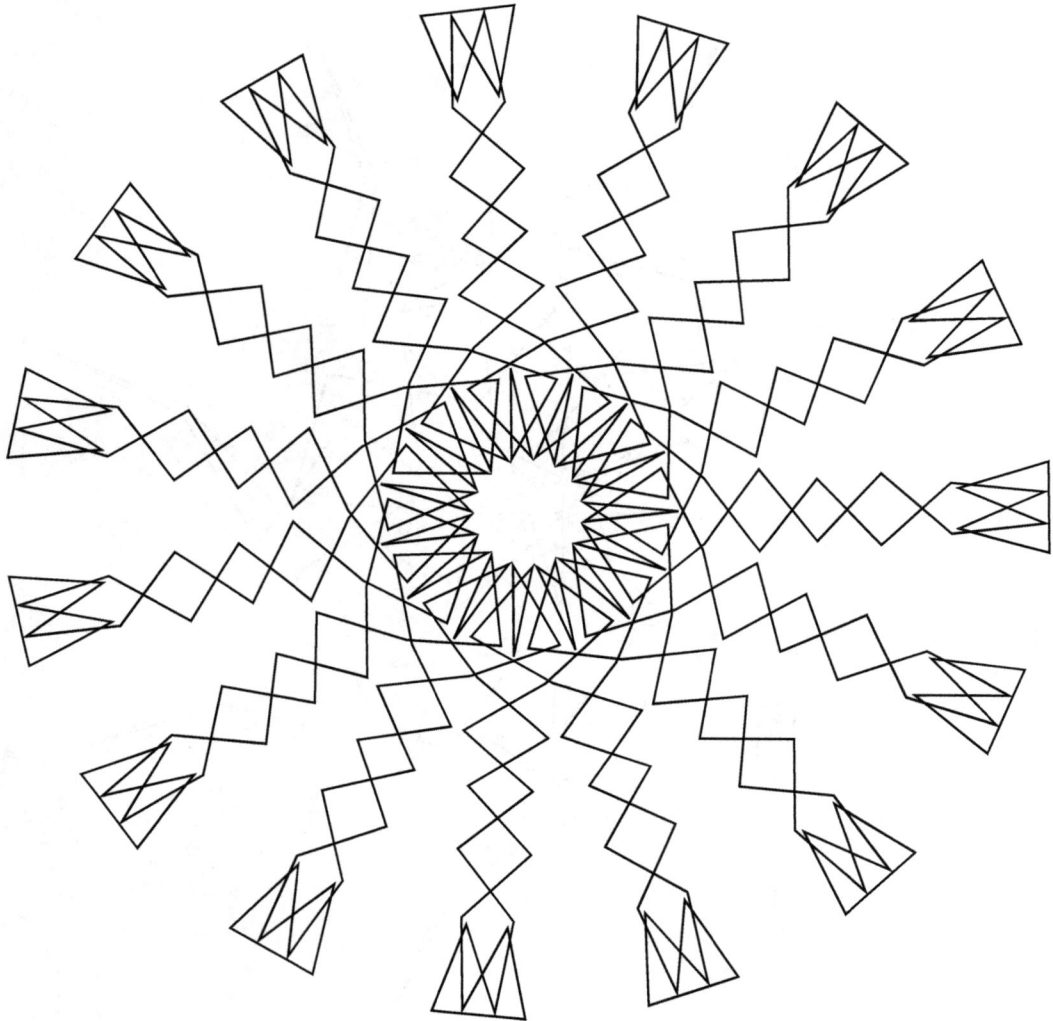

Figure 2.59 AAC [105: 0 79 40 95]

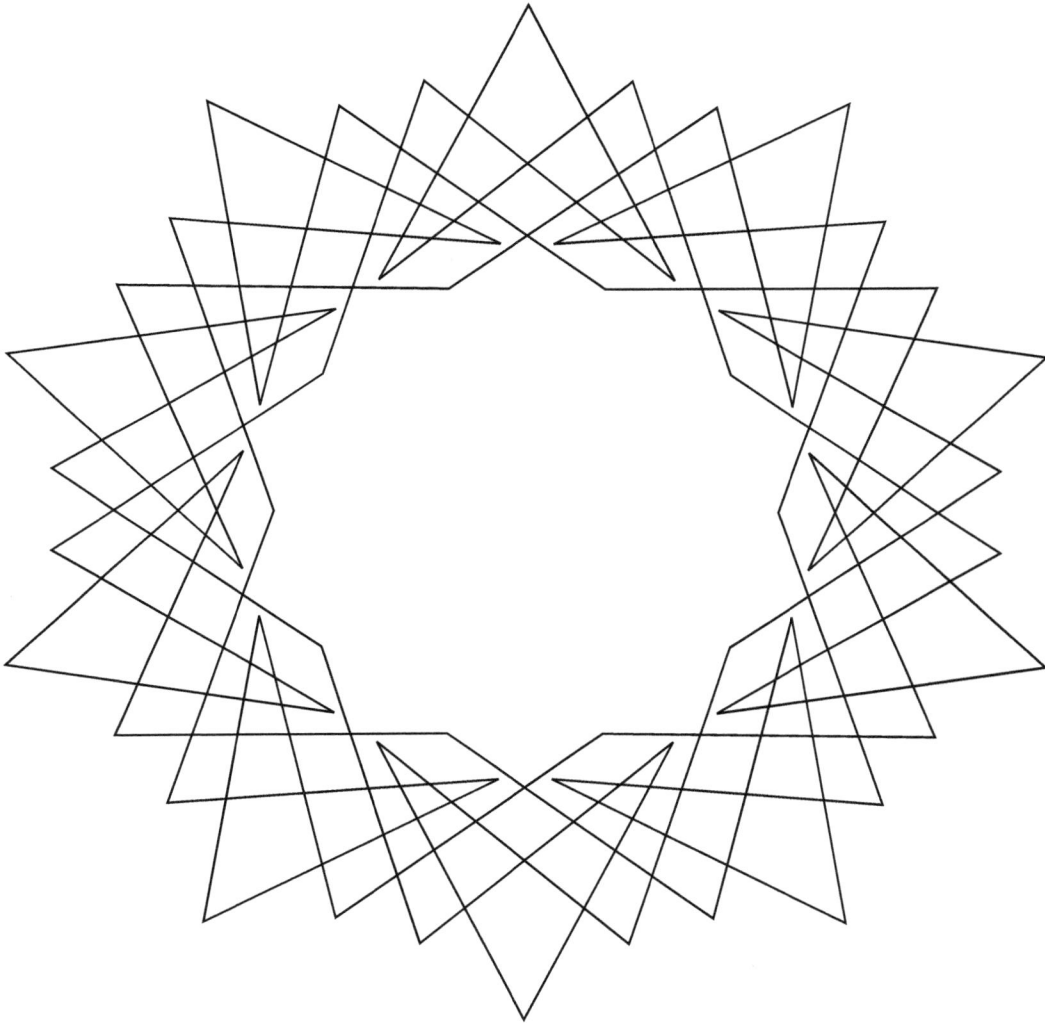

Figure 2.60   AAC [360: 0 247 270 60]

3 Accelerating Angle Curve Generation

This chapter describes accelerating angle curve generation (AACG). The chapter begins with a description of the coordinate systems used to generate and display an accelerating angle curve (AAC). Two examples of AACG are then examined in detail. The first example involves a closed AAC, and the second example involves an open AAC. An AACG algorithm is then provided. Finally, the format of the AAC signature is described.

3.1 AACG coordinate systems

An AAC is first generated in the raw coordinate system, shown in Figure 3.1 on the next page. The origin (0 0) is located toward the lower left corner of the reader's visual field. The x coordinates increase toward the right side of the reader's visual field, and the y coordinates increase toward the top of the reader's visual field. The x and y axis unit is called the *arbitrary length unit* (ALU). All AAC data in this book are presented relative to the raw coordinate system.

The angles employed in AACG, a, are measured from the positive x axis and increase in the counter-clockwise direction. The unit of a is specified as an integral number of angle units per circle, A; each unit is called a *slice*. For example, if A is 360, then one slice equals one degree. If A is 720, then a slice is half a degree. If A is 180, then there are two degrees per slice. In general, A is an integer greater than 1.

The AAC is transformed into the display coordinate system for presentation to the reader. This transformation preserves the AAC's orientation and aspect ratio; the AAC is merely scaled in x and y. The x and y axis unit is the pixel. The AACs in this book appear to have different line segment lengths due to the differing scale factors used in the raw-to-display coordinate system transformations. All AACs were first generated in the raw coordinate system using a line segment length of 10 ALUs.

65

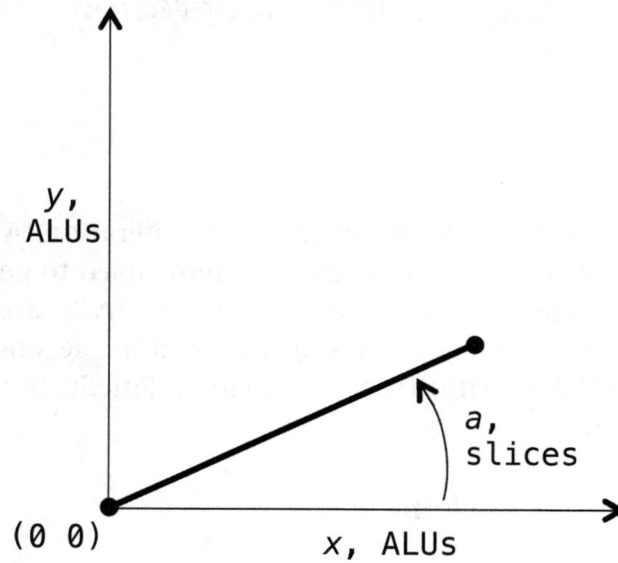

Figure 3.1 Raw coordinate system

3.2   Closed AACG example

Figure 3.2 on the facing page shows an AAC consisting of 16 line segments, numbered in the order of generation from 0 to 15. Each line segment s has start point $(pt[s].x\ pt[s].y)$ and end point $(pt[s+1].x\ pt[s+1].y)$, except for line segment 15, which has end point $(pt[0].x\ pt[0].y)$.

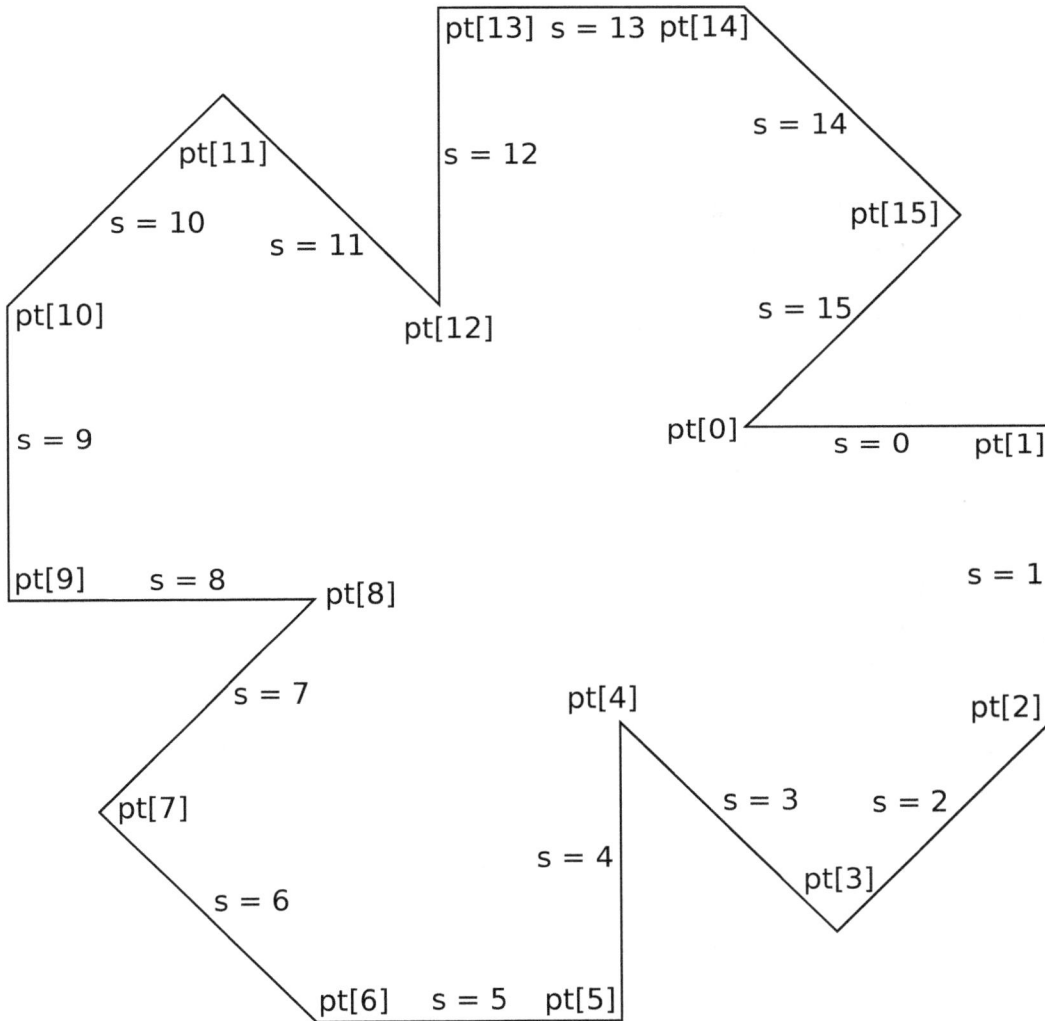

Figure 3.2 A simple closed AAC

Listing 3.1 shows the point and angle data for the AAC. The number of slices per circle, A, is 360, so one slice equals one degree. Each line of the listing describes an AAC line segment s by its start point, $pt[s]$, and its angle of orientation, called the *base angle*, a_0. The line segment length, 10 ALUs, is implicit and constant. Also associated with each line segment are its *forward difference angles*, a_1, a_2, and a_3. Forward difference angles a_4, a_5, a_6, and so on are all 0.

Listing 3.1

```
pt[ 0] = ( 10.000   10.000); a[0] ... a[3] =   0 270  45 270
pt[ 1] = ( 20.000   10.000); a[0] ... a[3] = 270 315 315 270
pt[ 2] = ( 20.000    0.000); a[0] ... a[3] = 225 270 225 270
pt[ 3] = ( 12.929   -7.071); a[0] ... a[3] = 135 135 135 270
pt[ 4] = (  5.858    0.000); a[0] ... a[3] = 270 270  45 270
pt[ 5] = (  5.858  -10.000); a[0] ... a[3] = 180 315 315 270
pt[ 6] = ( -4.142  -10.000); a[0] ... a[3] = 135 270 225 270
pt[ 7] = (-11.213   -2.929); a[0] ... a[3] =  45 135 135 270
pt[ 8] = ( -4.142    4.142); a[0] ... a[3] = 180 270  45 270
pt[ 9] = (-14.142    4.142); a[0] ... a[3] =  90 315 315 270
pt[10] = (-14.142   14.142); a[0] ... a[3] =  45 270 225 270
pt[11] = ( -7.071   21.213); a[0] ... a[3] = 315 135 135 270
pt[12] = ( -0.000   14.142); a[0] ... a[3] =  90 270  45 270
pt[13] = ( -0.000   24.142); a[0] ... a[3] =   0 315 315 270
pt[14] = ( 10.000   24.142); a[0] ... a[3] = 315 270 225 270
pt[15] = ( 17.071   17.071); a[0] ... a[3] = 225 135 135 270

pt[16] = ( 10.000   10.000); a[0] ... a[3] =   0 270  45 270
pt[17] = ( 20.000   10.000); a[0] ... a[3] = 270 315 315 270
pt[18] = ( 20.000    0.000); a[0] ... a[3] = 225 270 225 270
pt[19] = ( 12.929   -7.071); a[0] ... a[3] = 135 135 135 270
```

The AACG algorithm begins with the line segment 0 start point, $pt[0]$, line segment 0 angles, $a_0 - a_3$, and the line segment length, L. The line segment 1 start point, $pt[1]$, is calculated as a displacement of distance $L = 10$ ALUs from the line segment 0 start point in the direction specified by the line segment 0 base angle, a_0, which is 0 slices. $pt[1]$ is (20 10) ALUs. The start point of line segment 1 is also the end point of line segment 0.

The line segment 1 base angle, a_0', is then calculated by adding the line segment 0 *first order forward difference angle*, $a_1 = 270$, to the line segment 0 base angle, $a_0 = 0$, modulo $A = 360$. The result is $a_0' = 270$ slices. The line segment 1 first order forward difference angle, a_1', is then calculated by adding the line segment 0 *second order forward difference angle*, $a_2 = 45$, to the line segment 0 first order forward difference angle, $a_1 = 270$, modulo $A = 360$. The result is $a_1' = 315$ slices. The line segment 1 second and third order forward difference angles, a_2' and a_3', respectively, are calculated in a similar manner. These calculations can be summarized as:

$$a_0' = (a_0 + a_1) \bmod A = (0 + 270) \bmod 360 = 270$$
$$a_1' = (a_1 + a_2) \bmod A = (270 + 45) \bmod 360 = 315$$
$$a_2' = (a_2 + a_3) \bmod A = (45 + 270) \bmod 360 = 315$$
$$a_3' = (a_3 + a_4) \bmod A = (270 + 0) \bmod 360 = 270.$$

The AACG algorithm then calculates the line segment 2 start point, $pt[2]$, as a displacement of distance $L = 10$ ALUs from the line segment 1 start point, $pt[1]$, in the direction specified by the line segment 1 base angle, a_0', which is 270 slices. $pt[2]$ is (20 0) ALUs. As before, the start point of line segment 2 is also the end point of line segment 1.

The base and difference angles for line segment 2 are then calculated as:
$$a_0'' = (a_0' + a_1') \bmod A = (270 + 315) \bmod 360 = 225$$
$$a_1'' = (a_1' + a_2') \bmod A = (315 + 315) \bmod 360 = 270$$
$$a_2'' = (a_2' + a_3') \bmod A = (315 + 270) \bmod 360 = 225$$
$$a_3'' = (a_3' + a_4') \bmod A = (270 + 0) \bmod 360 = 270.$$

This alternation between generating the line segment s start point (which is also the line segment $s - 1$ end point) and generating the line segment s angles continues for line segments 3 through 15. The start point and angles of line segment 16 are identical to the start point and angles of line segment 0. Therefore, generating line segments 16 and beyond retraces the AAC already defined by line segments 0 through 15. For this reason, the AAC is declared *closed* at line segment 15. Listing 3.1 provides the start points and angles for line segments 16 through 19 to show that generating line segments beyond line segment 15 indeed retraces the AAC beginning at line segment 0.

The general rule for calculating a base or difference angle for line segment s from the base or difference angles for line segment $s - 1$ is

$$a'_n = (a_n + a_{n+1}) \bmod A$$

The order n of the highest-order non-zero line segment 0 angle is called the *AAC order*, N. In this example, the AAC order is 3.

3.3   Open AACG example

Figure 3.3 shows an AAC consisting of 21 line segments. Line segments 2 and 3, 9 and 10, and 16 and 17 are collinear, so open circles are used to show the location of the points joining these line segments. The portion of the AAC between points 0 and 7 is called the *fundamental subcurve*. The fundamental subcurve repeats twice: between points 7 and 14 and between points 14 and 21.

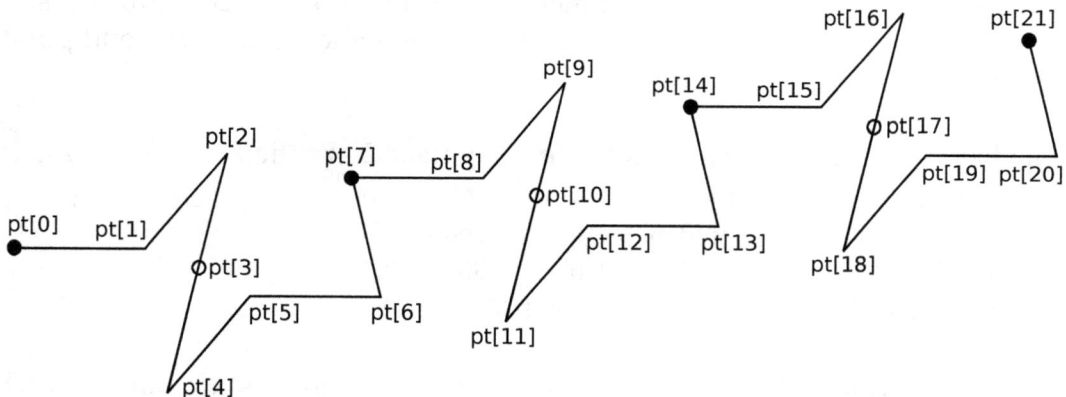

Figure 3.3   A simple open AAC

Listing 3.2 shows the point and angle data for this order 2 AAC. Every 7th line segment the angles repeat and the start point is displaced by a fixed distance in the xy plane.

Listing 3.2

```
pt[ 0] = ( 10.000  10.000); a[0] ... a[2] =   0   1   3
pt[ 1] = ( 20.000  10.000); a[0] ... a[2] =   1   4   3
pt[ 2] = ( 26.235  17.818); a[0] ... a[2] =   5   0   3
pt[ 3] = ( 24.010   8.069); a[0] ... a[2] =   5   3   3
pt[ 4] = ( 21.784  -1.680); a[0] ... a[2] =   1   6   3
pt[ 5] = ( 28.019   6.138); a[0] ... a[2] =   0   2   3
pt[ 6] = ( 38.019   6.138); a[0] ... a[2] =   2   5   3

pt[ 7] = ( 35.794  15.887); a[0] ... a[2] =   0   1   3
pt[ 8] = ( 45.794  15.887); a[0] ... a[2] =   1   4   3
pt[ 9] = ( 52.029  23.706); a[0] ... a[2] =   5   0   3
pt[10] = ( 49.804  13.956); a[0] ... a[2] =   5   3   3
pt[11] = ( 47.579   4.207); a[0] ... a[2] =   1   6   3
pt[12] = ( 53.814  12.025); a[0] ... a[2] =   0   2   3
pt[13] = ( 63.814  12.025); a[0] ... a[2] =   2   5   3

pt[14] = ( 61.588  21.775); a[0] ... a[2] =   0   1   3
pt[15] = ( 71.588  21.775); a[0] ... a[2] =   1   4   3
pt[16] = ( 77.823  29.593); a[0] ... a[2] =   5   0   3
pt[17] = ( 75.598  19.844); a[0] ... a[2] =   5   3   3
pt[18] = ( 73.373  10.094); a[0] ... a[2] =   1   6   3
pt[19] = ( 79.608  17.913); a[0] ... a[2] =   0   2   3
pt[20] = ( 89.608  17.913); a[0] ... a[2] =   2   5   3

pt[21] = ( 87.383  27.662); a[0] ... a[2] =   0   1   3
```

The number of slices per circle, A, is 7. Each slice is exactly $\frac{360}{7}$ degrees ($\sim$51.4). Table 3.1 shows how to convert the angles in Listing 3.2 from slices to degrees, rounded to the nearest tenth of a degree.

Table 3.1 Converting $A = 7$ slices into degrees

Slices	Degrees
0	0.0
1	51.4
2	102.9
3	154.3
4	205.7
5	257.1
6	308.6

The AACG algorithm begins with the line segment 0 start point, $pt[0]$, line segment 0 angles, $a_0 - a_2$, and the line segment length, L. The line segment 1 start point, $pt[1]$, is calculated as a displacement of distance $L = 10$ ALUs from the line segment 0 start point in the direction specified by the line segment 0 base angle, a_0, which is 0 slices. $pt[1]$ is (20 10) ALUs.

The base and difference angles for line segment 1 are then calculated as:
$a_0' = (a_0 + a_1) \bmod A = (0 + 1) \bmod 7 = 1$
$a_1' = (a_1 + a_2) \bmod A = (1 + 3) \bmod 7 = 4$
$a_2' = (a_2 + a_3) \bmod A = (3 + 0) \bmod 7 = 3.$

The AACG algorithm then calculates the line segment 2 start point, $pt[2]$, as a displacement of distance $L = 10$ ALUs from the line segment 1 start point, $pt[1]$, in the direction specified by the line segment 1 base angle, a_0', which is 1 slice ($\sim$51.4 degrees). $pt[2]$ is (26.235 17.818) ALUs.

The base and difference angles for line segment 2 are then calculated as:
$a_0'' = (a_0' + a_1') \bmod A = (1 + 4) \bmod 7 = 5$
$a_1'' = (a_1' + a_2') \bmod A = (4 + 3) \bmod 7 = 0$
$a_2'' = (a_2' + a_3') \bmod A = (3 + 0) \bmod 7 = 3.$

This process continues for line segments 3 through 20. There are a total of 21 line segments in the AAC.

The line segment 7 base and difference angles are identical to the line segment 0 base and difference angles, but the line segment 7 start point is different than the line segment 0 start point. Therefore, the subcurve between points 0 and 7 is the fundamental subcurve and repeats between points 7 and 14 with displacement from point 0 in the x and y directions of $(pt[7].x - pt[0].x) = \sim 25.794$ and $(pt[7].y - pt[0].y) = \sim 5.887$ ALUs, respectively. Similarly, the line segment 14 angles are identical to the line segment 7 angles, so the fundamental subcurve repeats again between points 14 and 21 with the same displacement from point 7.

Generating line segments beyond line segment 20 in multiples of 7 creates an AAC containing additional repetitions of the fundamental subcurve. If the AAC has S' line segments, where S' is a multiple of 7, it will contain $\frac{S'}{7} - 1$ repetitions of the fundamental subcurve.

3.4   AACG algorithm

3.4.1   AACG function

Figure 3.4 provides pseudocode for the `AACG` function, which generates an AAC in the raw coordinate system.

AACG($N, S, pt[\ ], a[\ ], L, A$**)**
  Input:
  - AAC order, N
  - number of line segments in the AAC, S
  - line segment 0 start point, $pt[0]$, ALUs
  - line segment 0 angles, $a[\ ] =< a_0, a_1, a_2, \cdots, a_N >$, slices
  - line segment length, L, ALUs
  - number of slices per circle, A

  Output:
  - line segment 1 to $(S-1)$ start points, $pt[1], pt[2], \cdots, pt[S-1]$, ALUs
  - line segment $(S-1)$ angles, $a[\ ] =< a_0, a_1, a_2, \cdots, a_N >$, slices

```
// calculate conversion factor: slices to radians
```
  $f = (2 * \pi) \ / \ A$
```
// generate AAC line segments 1 through S - 1
```
  for $s = 1$ **to** $(S-1)$
```
    // calculate line segment s start point
```
    $pt[s].x = pt[s-1].x + L * \cos(a_0 * f)$
    $pt[s].y = pt[s-1].y + L * \sin(a_0 * f)$
```
    // calculate line segment s angles
```
    for $n = 0$ **to** $(N-1)$
      $a_n = (a_n + a_{n+1}) \bmod A$
    end
  end
end

Figure 3.4   Accelerating angle curve generation function

The function begins by calculating a conversion factor f between slices and radians, because the cos and sin functions accept an angle with radian units as input. The function then generates $S-1$ line segments. Each iteration of the line segment generation loop involves:

· calculating the start point $pt[s]$ of line segment s as a displacement of L ALUs from the previous line segment's start point $pt[s-1]$ along a line with direction specified by the previous line segment's base angle a_0.

· calculating base and difference angles $a_0 \cdots a_{N-1}$ for line segment s.

Since all difference angles of order $n > N$ are zero, all line segments generated have the same Nth order difference angle, a_N.

3.4.2 Determining the number of line segments in an AAC

When the number of line segments S in the AAC is not given but must instead be determined, the AACG algorithm can be easily modified to generate line segments and angles until a line segment s is found with angles equal to the line segment 0 angles. If the start point of the new line segment, $pt[s]$, equals the start point of line segment 0, $pt[0]$, then the AAC has closed, and the number of line segments in the AAC, S, is s. Otherwise, the AAC is open, and by further iterating the line segment generation loop the number of line segments in the AAC can be set to any value desired.

3.5 AAC signature

The AACs in this book are informally identified using a subset of the AACG algorithm parameters called the *AAC signature*. The format of the AAC signature is

$$[A : a_0\ a_1\ a_2\ \cdots\ a_N],$$

where A is the number of slices per circle and $a_0\ a_1\ a_2\ \cdots\ a_N$ are the line segment 0 angles. For example, the AAC in Figure 3.2 on page 67 has signature [360: 0 270 45 270] and the AAC in Figure 3.3 on page 70 has signature [7: 0 1 3].

The AACG algorithm parameters included in the AAC signature are sufficient to qualitatively define the AAC. The AAC start point $pt[0]$ is omitted because changing this parameter merely changes the location of the AAC in the xy plane. The AAC line segment length, L, is omitted because changing this parameter merely changes the overall size of the AAC in the raw coordinate system but does not change its structure. The number of line segments in the AAC, S, is omitted because for a closed AAC it is understood that S is whatever value is necessary to close the curve. For an open AAC, changing S changes the overall size of the AAC in the raw coordinate system but does not change its structure, which consists of repetitions of its fundamental subcurve.

4 Accelerating Angle Curve Symmetry

This chapter is an introduction to the two types of closed AAC symmetry examined in this book: rotational and reflectional. The first section discusses AAC rotational symmetry, and the second section discusses AAC reflectional symmetry.

4.1 AAC rotational symmetry

An AAC has an *angle a rotational symmetry* if rotating the entire AAC counterclockwise by a around the AAC's centroid results in a rotated AAC tracing the same path in the same direction in the xy plane as the original AAC. A rotational symmetry where the rotation angle a is less than a full circle is called *nontrivial*, whereas rotation by a full circle is called *trivial*.[1] This section contains two examples in which the number of rotational symmetries of an AAC are determined.

4.1.1 Rotational symmetry example 1

Figure 4.1 shows the order 6 AAC [4: 0 1 0 2 1 0 2]. The 16 points in the AAC are numbered in generation order from 0 to 15.

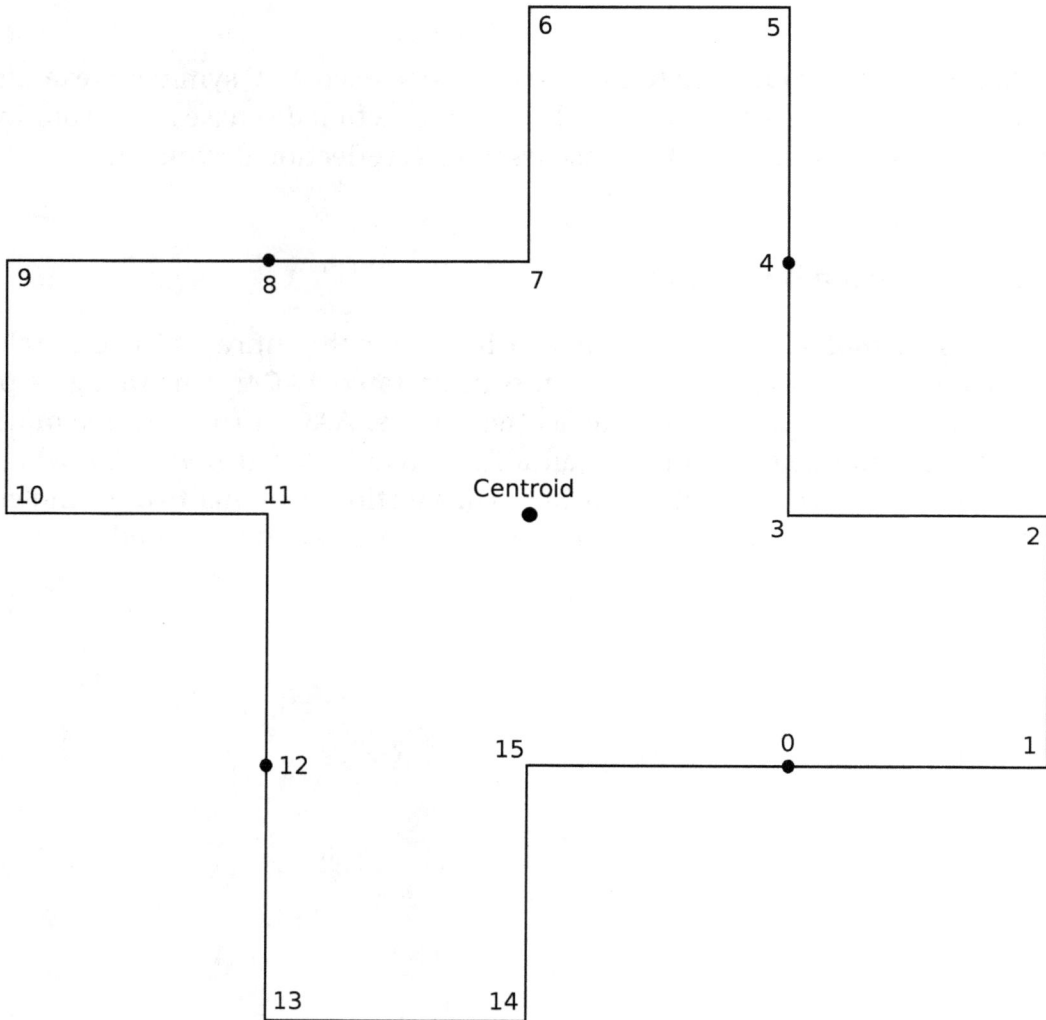

Figure 4.1   An AAC with four rotational symmetries

The AAC's rotational symmetry count can be determined by noting that:

· when the AAC is rotated counter-clockwise around its centroid through one slice (90 degrees), point 0 maps onto point 4, point 1 maps onto point 5, and so on up to point 15, which maps onto point 3. The rotated AAC traces the same path in the same direction in the xy plane as the original AAC.

· when the AAC is rotated counter-clockwise around its centroid through two slices (180 degrees), point 0 maps onto point 8, point 1 maps onto point 9, and so on up to point 15, which maps onto point 7. The rotated AAC traces the same path in the same direction in the xy plane as the original AAC.

· when the AAC is rotated counter-clockwise around its centroid through three slices (270 degrees), point 0 maps onto point 12, point 1 maps onto point 13, and so on up to point 15, which maps onto point 11. The rotated AAC traces the same path in the same direction in the xy plane as the original AAC.

· when the AAC is rotated counter-clockwise around its centroid through four slices (360 degrees), point 0 maps onto point 0, point 1 maps onto point 1, and so on up to point 15, which maps onto point 15. The rotated AAC traces the same path in the same direction in the xy plane as the original AAC.

· no other rotations of the AAC around its centroid result in a rotated AAC that traces the same path in the same direction in the xy plane as the original AAC. Rotational symmetries corresponding to rotations greater than a circle are excluded from the AAC's rotational symmetry count.

Therefore, the AAC has four rotational symmetries.

4.1.2 Rotational symmetry example 2

Figure 4.2 shows the order 2 AAC [35: 0 2 14]. The 35 points in the AAC are numbered in generation order from 0 to 34.

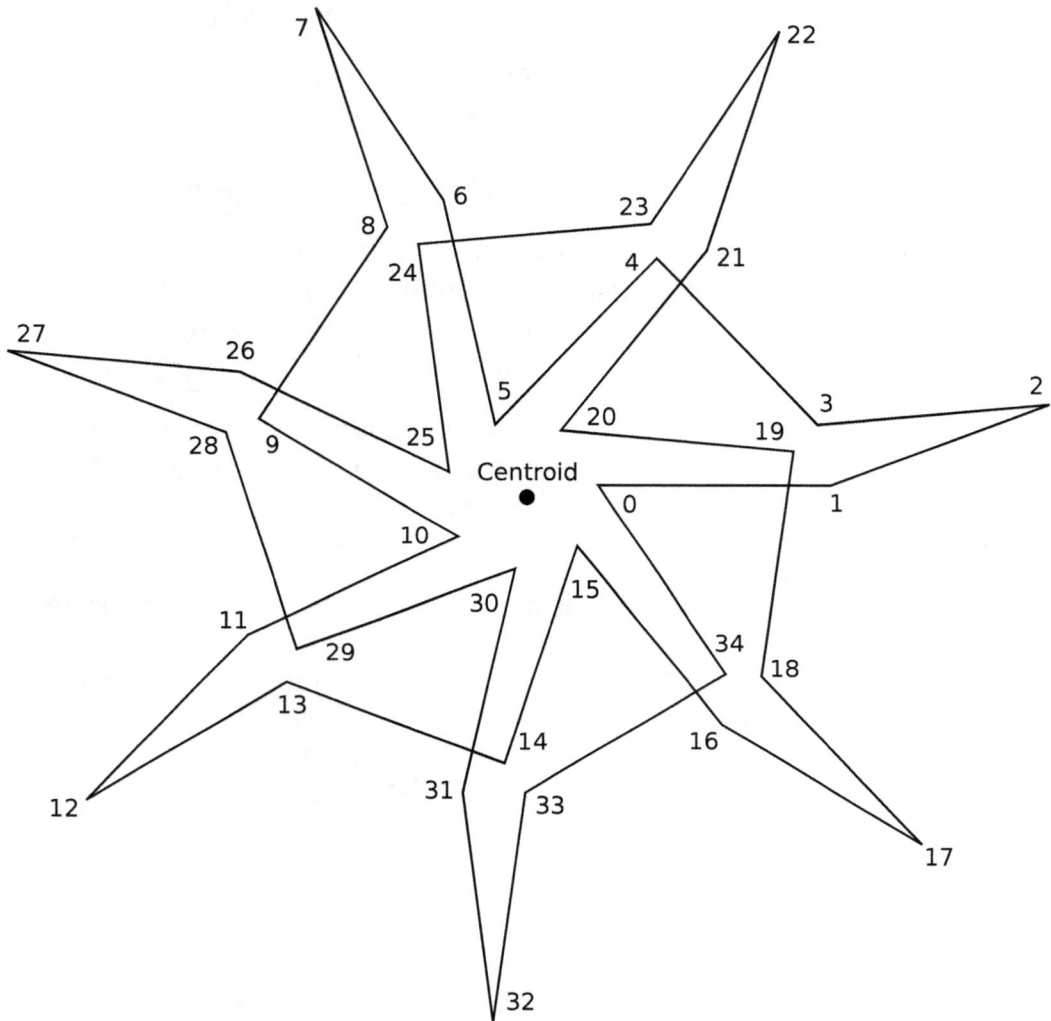

Figure 4.2 An AAC with seven rotational symmetries

The AAC's rotational symmetry count can be determined by noting that when the AAC is rotated counter-clockwise around its centroid through 5, 10, 15, 20, 25, 30, or 35 slices (corresponding to approximately 51.4, 102.9, 154.3, 205.7, 257.1, 308.6, or 360 degrees, respectively), the rotated AAC traces the same path in the same direction in the xy plane as the original AAC. No other rotations of the AAC around its centroid have this property. Therefore, the AAC has seven rotational symmetries.

4.2 AAC reflectional symmetry

A line in the xy plane is a *reflection axis* of an AAC if rotating the entire AAC 180 degrees around the line through the third dimension z results in a rotated AAC tracing the same path in the xy plane as the original AAC, but in the opposite direction [7, p. viii]. This section contains two examples in which the number of reflectional symmetries of an AAC are determined.

4.2.1 Reflectional symmetry example 1

Figure 4.3 shows the order 3 AAC [360: 0 20 270 210]. The 24 points in the AAC are numbered in generation order from 0 to 23.

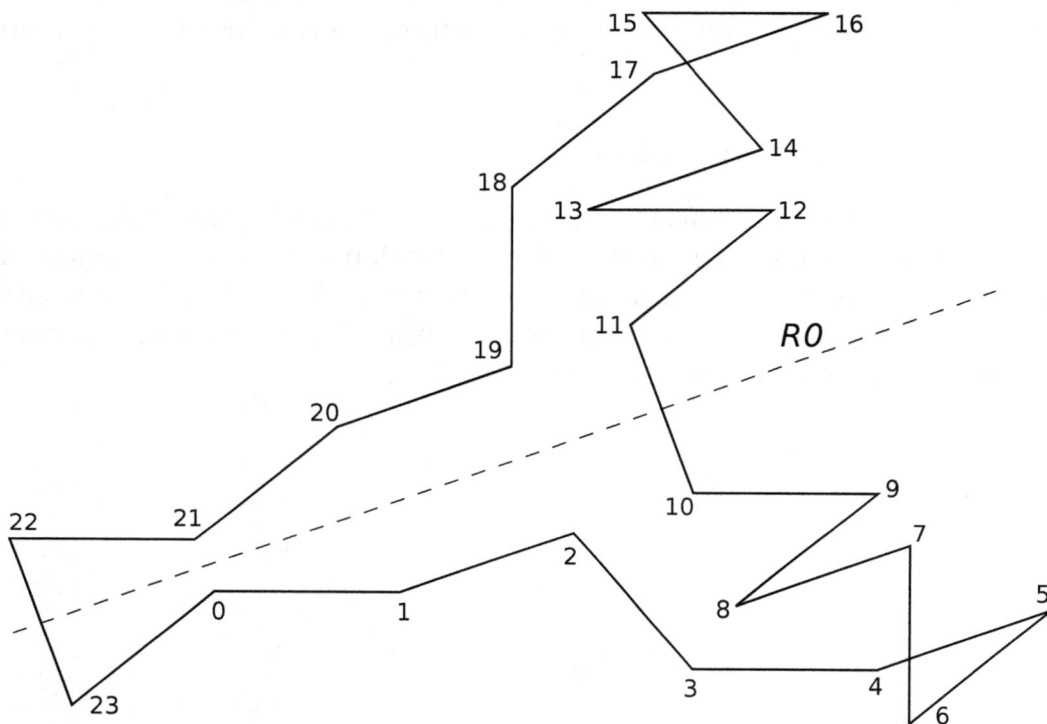

Figure 4.3   An AAC with one reflectional symmetry

When the AAC is rotated 180 degrees around its reflection axis *R0* through the *z* dimension, point 22 exchanges places with point 23, point 21 exchanges places with point 0, point 20 exchanges places with point 1, and so on up to point 11, which exchanges places with point 10. The rotated AAC traces the same path in the *xy* plane as the original AAC, but in the opposite direction. There are no other lines in the *xy* plane which qualify as reflection axes. Therefore, the AAC has one reflectional symmetry.

4.2.2 Reflectional symmetry example 2

Figure 4.4 shows the order 3 AAC [9: 0 8 7 4]. The 27 points in the AAC are numbered in generation order from 0 to 26. Points 2, 11, and 20 coincide.

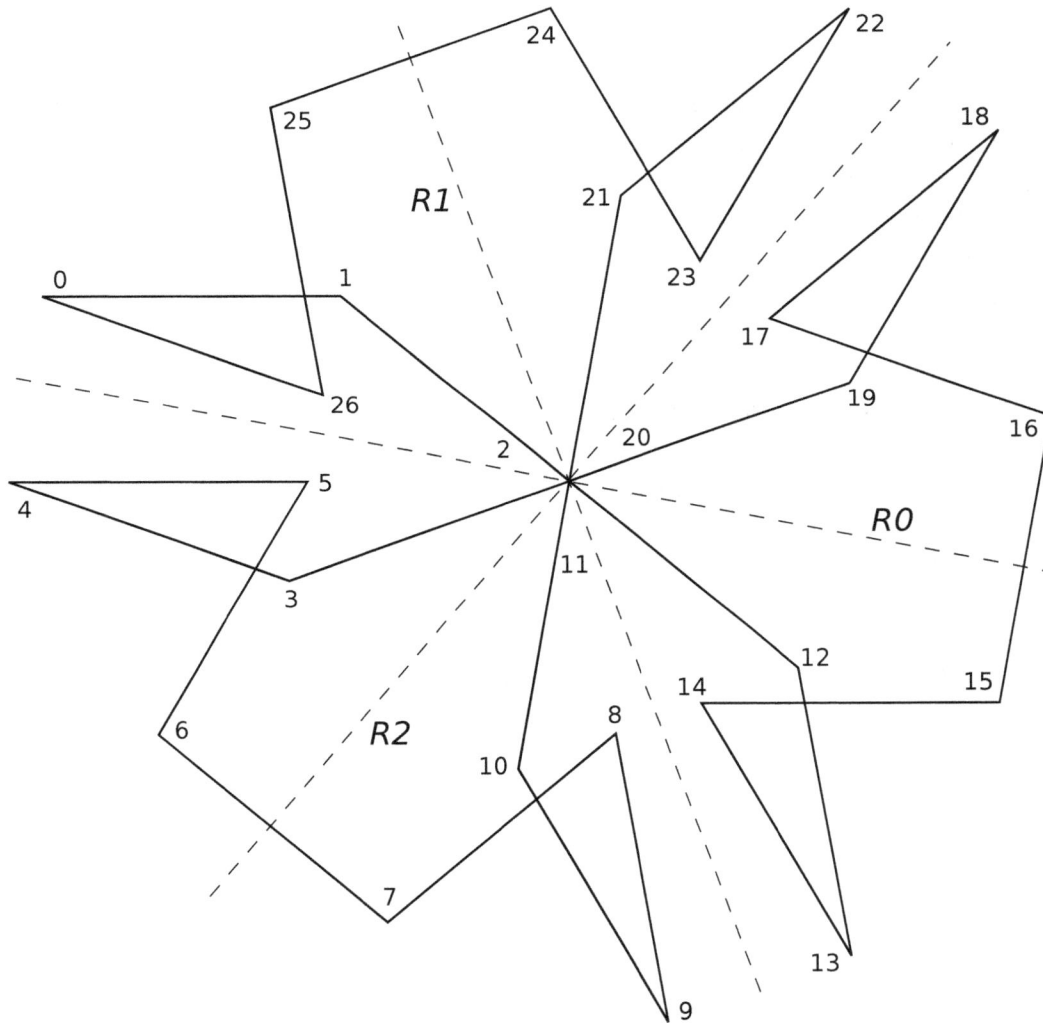

Figure 4.4 An AAC with three reflectional symmetries

When the AAC is rotated 180 degrees around its reflection axis $R0$ through the z dimension, point 15 exchanges places with point 16, point 14 exchanges places with point 17, point 13 exchanges places with point 18, and so on up to point 3, which exchanges places with point 1. Point 2 does not exchange places with another point. The rotated AAC traces the same path in the xy plane as the original AAC, but in the opposite direction.

When the AAC is rotated 180 degrees around its reflection axis $R1$ through the z dimension, point 24 exchanges places with point 25, point 23 exchanges places with point 26, point 22 exchanges places with point 0, and so on up to point 12, which exchanges places with point 10. Point 11 does not exchange places with another point. The rotated AAC traces the same path in the xy plane as the original AAC, but in the opposite direction.

When the AAC is rotated 180 degrees around its reflection axis $R2$ through the z dimension, point 6 exchanges places with point 7, point 5 exchanges places with point 8, point 4 exchanges places with point 9, and so on up to point 21, which exchanges places with point 19. Point 20 does not exchange places with another point. The rotated AAC traces the same path in the xy plane as the original AAC, but in the opposite direction.

There are no other lines in the xy plane which qualify as reflection axes. Therefore, the AAC has three reflectional symmetries.

Notes

[1]Should the trivial rotational symmetry *really* be considered a property of an AAC? For example, would the AAC in Figure 5.9 on page 114 be more accurately characterized as *completely lacking* rotational symmetry rather than having only *trivial* rotational symmetry? This book includes rotation by a full circle as a rotational symmetry of an AAC for the following reasons:

- The rotational symmetry analysis is simplified.

- For an AAC having nontrivial rotational symmetry, it is often easy to find its total number of rotational symmetries by counting the number of times a distinctive feature occurs in the AAC. For example, in Figure 4.2 on page 80 the AAC has seven distinctive outward-pointed spikes and also has seven rotational symmetries.

- There is a long historical precedent of including the trivial rotational symmetry as a rotational symmetry of an object. This is due in part to the use of *group theory* as a theoretical foundation of symmetry analysis [8, p. 42].

5 Predicting Accelerating Angle Curve Rotational Symmetry

This chapter describes how to predict the number of rotational symmetries (PRTS) of a closed AAC given only its signature. The first section provides a PRTS algorithm. The algorithm consists of a main routine and two subroutines. The following seven sections provide examples of PRTS algorithm operation. The first three examples are order 1 AACs, the fourth example is an order 2 AAC, and the last three examples are order 3 AACs. Each example is divided into four parts: (1) a graph of the AAC; (2) a listing containing the AAC's point and angle data; (3) a trace of PRTS algorithm operation; and (4) a discussion of the trace.

5.1   PRTS algorithm

This section describes the PRTS algorithm main routine, PRTS, and its two subroutines, EFDM and LCM.

5.1.1   PRTS function

Figure 5.1 on the next page provides pseudocode for the PRTS function. The function inputs are the number of slices per circle, the AAC order, and the line segment 0 angles. The function returns the predicted number of AAC rotational symmetries.

PRTS$(A, N, a[\,])$
> **Input:**
>> • number of slices per circle, A
>> • AAC order, N
>> • line segment 0 angles, $a[\,] = <a_0, a_1, \cdots, a_N>$, slices
>
> **Output:**
>> • predicted number of AAC rotational symmetries, $prts$
>
> **Working:**
>> • difference order, n
>> • number of iterations per $<a_n \cdots a_N>$ rollover, $nipr_n$
>> • value of angle a_n after $nipr_{n+1}$ iterations, a'_n
>> • change in angle a_n after $nipr_{n+1}$ iterations, da_n
>> • number of $<a_{n+1} \cdots a_N>$ rollovers per $<a_n \ldots a_N>$ rollover, nr_n

```
// initialize: one iteration per aN rollover
```
$nipr_N = 1$
```
// for difference orders n from (N − 1) to 0 ...
```
for $n = N - 1$ **to** $n = 0$
>```
>    // find the change in an after niprn+1 iterations
>```
> $a'_n = \text{EFDM}(A, N - n, <a_n \cdots a_N>, nipr_{n+1})$
> $da_n = (a'_n - a_n) \mod A$
>```
>    // find the number of <an+1···aN> rollovers per
>    //     <an...aN> rollover
>```
> **if** $da_n = 0$ **then**
>> $nr_n = 1$
>
> **else**
>> $nr_n = \text{LCM}(A, da_n) \,/\, da_n$
>
> **end**
>```
>    // find the number of iterations per <an···aN> rollover
>```
> $nipr_n = nipr_{n+1} * nr_n$

end
```
// conclude: return the predicted number of AAC
//     rotational symmetries
```
$prts = nr_0$
end

Figure 5.1   AAC rotational symmetry prediction function

5.1.2 EFDM function

Figure 5.2 provides pseudocode for the Extrapolate Forward Differences with Modulo (EFDM) subroutine. This function calculates the new value g_0' of base angle g_0 after t iterations of the base and forward difference angles $g = < g_0, g_1, \cdots g_f >$, where f is the number of forward differences. The function consists of *Newton's Forward Difference Formula* with an appended modulo A operation.

EFDM(A, f, g, t)
   Input:
     • number of slices per circle, A
     • number of forward difference angles, f
     • base and forward difference angles, $g = < g_0, g_1, \cdots, g_f >$, slices
     • number of iterations, t
   Output:
     • base angle g_0 after t iterations, g_0', slices

$$g_0' = (\ g_0 + g_1\frac{(t-0)}{1!} + g_2\frac{(t-0)(t-1)}{2!} + g_3\frac{(t-0)(t-1)(t-2)}{3!} + \cdots$$
$$g_f\frac{(t-0)(t-1)(t-2)\cdots(t-(f-1))}{f!}\ )\quad \mathrm{mod}\ A$$

end

Figure 5.2   Extrapolate forward differences with modulo function

The base angle g_0 employed by the EFDM function may be different than the base angle a_0 employed by the PRTS function. For example, if function PRTS calls function EFDM with parameter $g = < a_2, a_3, a_4 >$, then $g_0 = a_2$, $g_1 = a_3$, and $g_2 = a_4$.

5.1.3 LCM function

The least common multiple function LCM(A, da_n) returns the smallest integer that is evenly divisible by integers A and da_n. The least common multiple ranges from a minimum of the greater of A and da_n to a maximum of the product of A and da_n. Algorithms for finding the least common multiple are widely available and thus a LCM function is not provided in this book.

Chapter 5

5.2   PRTS example 1

5.2.1   Graph

Figure 5.3 shows an order 1 AAC with five rotational symmetries.

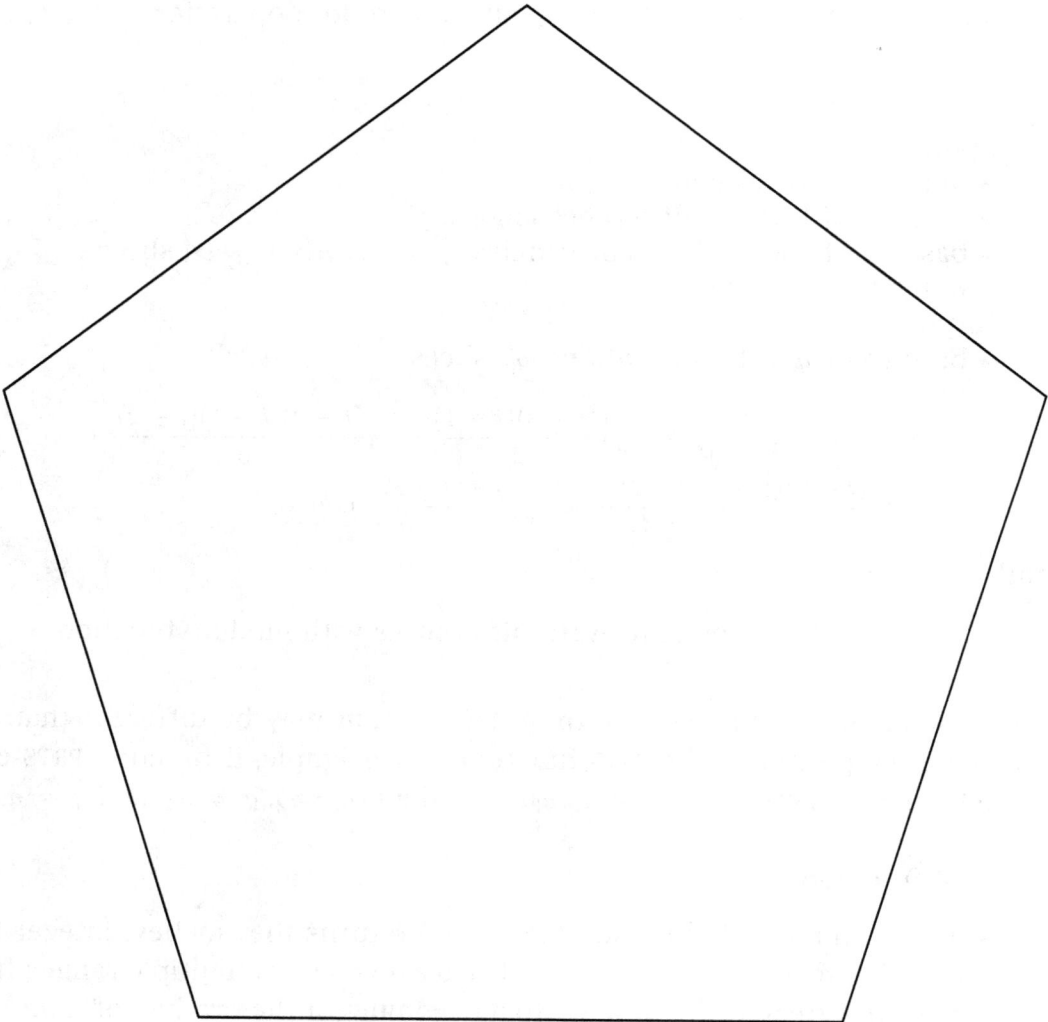

Figure 5.3   AAC [35: 0 7]

90

5.2.2 Point and angle data

Listing 5.1 shows the point and angle data for the AAC.

Listing 5.1

```
pt[ 0] = (10.000 10.000); a[0] a[1] =    0    7
pt[ 1] = (20.000 10.000); a[0] a[1] =    7    7
pt[ 2] = (23.090 19.511); a[0] a[1] =   14    7
pt[ 3] = (15.000 25.388); a[0] a[1] =   21    7
pt[ 4] = ( 6.910 19.511); a[0] a[1] =   28    7

pt[ 5] = (10.000 10.000); a[0] a[1] =    0    7
```

5.2.3 Trace

The PRTS algorithm predicts the number of rotational symmetries of AAC $[A : a_0\ a_1] = [35 : 0\ 7]$ as follows:

$$- \text{ initialize } -$$

(1) set the number of iterations per a_1 rollover

$\qquad nipr_1 = 1$

$$- n = 0 -$$

(2) find the change in a_0 after $nipr_1$ iterations

$\qquad a_0' = \text{EFDM}(A, N - n, < a_0\ a_1 >, nipr_1) = \text{EFDM}(35, 1, < 0\ 7 >, 1) = 7$

$\qquad da_0 = (a_0' - a_0) \mod A = (7 - 0) \mod 35 = 7$

(3) find the number of a_1 rollovers per $< a_0\ a_1 >$ rollover

$\qquad nr_0 = \text{LCM}(A, da_0)/da_0 = \text{LCM}(35, 7)/7 = \frac{35}{7} = 5$

(4) find the number of iterations per $< a_0\ a_1 >$ rollover

$\qquad nipr_0 = nipr_1 * nr_0 = 1 * 5 = 5$

$$- \text{ conclude } -$$

(5) return the predicted number of AAC rotational symmetries

$\qquad prts = nr_0 = 5$

5.2.4 Discussion

Whenever the angles $< a_{n1} \cdots a_{n2} >$ at some line segment s are equal to those angles at line segment 0, the angles are said to have *rolled over* at line segment s. For example, in Listing 5.1 angle $a_1 = 7$ rolls over at line segments $s = 1, 2, 3$, and so on, while angles $< a_0 \ a_1 > = < 0 \ 7 >$ roll over at line segment $s = 5$. If Listing 5.1 were extended beyond line segment 5, it would be apparent that $< a_0 \ a_1 >$ also rolls over at line segments $s = 10, 15, 20$, and so on.

Step 1: angle $a_1 = 7$ rolls over every iteration, as can be seen in Listing 5.1, so the number of iterations per a_1 rollover is 1. In general, for an order N AAC angle a_N rolls over every iteration.

Step 2: the trace shows the general method of finding da_0, the change in a_0 after $nipr_1 = 1$ iteration, using the EFDM function. However, this method is unnecessarily complex for this AAC. Listing 5.1 shows that the change in a_0 after one iteration is simply a_1, which is 7.

Step 3: the trace shows the general method of finding nr_0, the number of a_1 rollovers per $< a_0 \ a_1 >$ rollover, using the LCM function. However, this method is unnecessarily complex for this AAC. With 35 slices per circle and a_0 increasing by $da_0 = 7$ slices ($360 * \frac{7}{35} = 72$ degrees) each time a_1 rolls over, $< a_0 \ a_1 >$ will roll over after $\frac{A}{da_0} = \frac{35}{7} = 5$ a_1 rollovers. This can be understood intuitively by examining Figure 5.3 on page 90.

Step 4: with $nipr_1 = 1$ iteration per a_1 rollover and $nr_0 = 5$ a_1 rollovers per $< a_0 \ a_1 >$ rollover, there will be $nipr_0 = 5$ iterations per $< a_0 \ a_1 >$ rollover.

Step 5: the AAC's predicted number of rotational symmetries is the predicted number of a_1 rollovers per $< a_0 \ a_1 >$ rollover, which is 5.

5.3 PRTS example 2

5.3.1 Graph

Figure 5.4 shows an order 1 AAC with seven rotational symmetries.

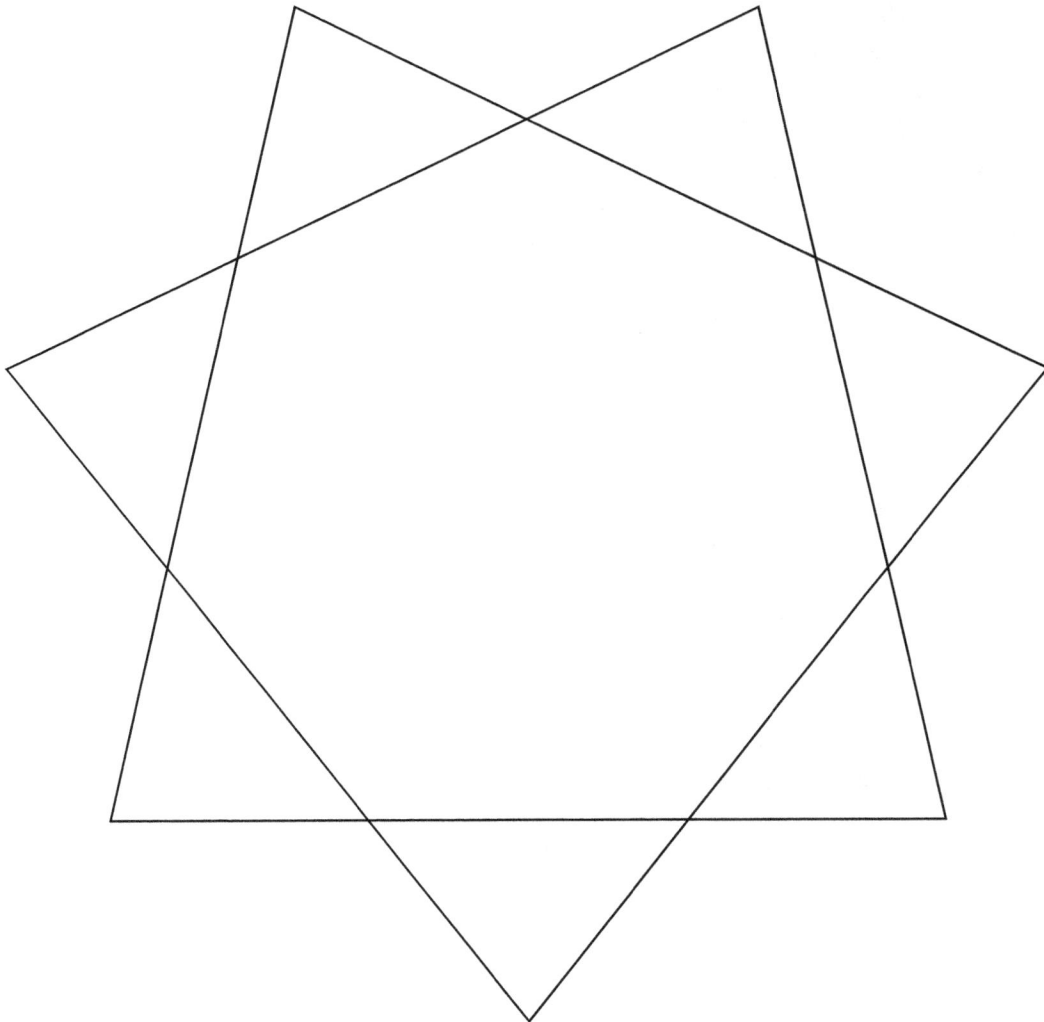

Figure 5.4 AAC [35: 0 10]

5.3.2   Point and angle data

Listing 5.2 shows the point and angle data for the AAC.

Listing 5.2

```
pt[ 0] = (10.000 10.000); a[0] a[1] =    0   10
pt[ 1] = (20.000 10.000); a[0] a[1] =   10   10
pt[ 2] = (17.775 19.749); a[0] a[1] =   20   10
pt[ 3] = ( 8.765 15.410); a[0] a[1] =   30   10
pt[ 4] = (15.000  7.592); a[0] a[1] =    5   10
pt[ 5] = (21.235 15.410); a[0] a[1] =   15   10
pt[ 6] = (12.225 19.749); a[0] a[1] =   25   10


pt[ 7] = (10.000 10.000); a[0] a[1] =    0   10
```

5.3.3   Trace

The PRTS algorithm predicts the number of rotational symmetries of AAC $[A : a_0\ a_1] = [35 : 0\ 10]$ as follows:

$$- \text{ initialize } -$$

(1) set the number of iterations per a_1 rollover

$\quad nipr_1 = 1$

$$- n = 0 -$$

(2) find the change in a_0 after $nipr_1$ iterations

$\quad a_0' = \text{EFDM}(A, N - n, < a_0\ a_1 >, nipr_1) = \text{EFDM}(35, 1, < 0\ 10 >, 1) = 10$

$\quad da_0 = (a_0' - a_0) \mod A = (10 - 0) \mod 35 = 10$

(3) find the number of a_1 rollovers per $< a_0\ a_1 >$ rollover

$\quad nr_0 = \text{LCM}(A, da_0)/da_0 = \text{LCM}(35, 10)/10 = 7$

(4) find the number of iterations per $< a_0\ a_1 >$ rollover

$\quad nipr_0 = nipr_1 * nr_0 = 1 * 7 = 7$

$$- \text{ conclude } -$$

(5) return the predicted number of AAC rotational symmetries

$\quad prts = nr_0 = 7$

5.3.4 Discussion

Step 3: this AAC differs from the previous AAC in that it is not possible to use the simple expression $\frac{A}{da_0}$ to find nr_0, the number of a_1 rollovers per $< a_0 \; a_1 >$ rollover, because $da_0 = 10$ does not divide evenly into $A = 35$. Instead, the full expression $\text{LCM}(A, da_0)/da_0$ must be used. Using the least common multiple of A and da_0 in the numerator of the quotient guarantees that the denominator da_0 always evenly divides the numerator. An examination of Figure 5.4 on page 93 and Listing 5.2 provides an intuitive basis for this approach.

5.4   PRTS example 3

5.4.1   Graph

Figure 5.5 shows an order 1 AAC with 35 rotational symmetries.

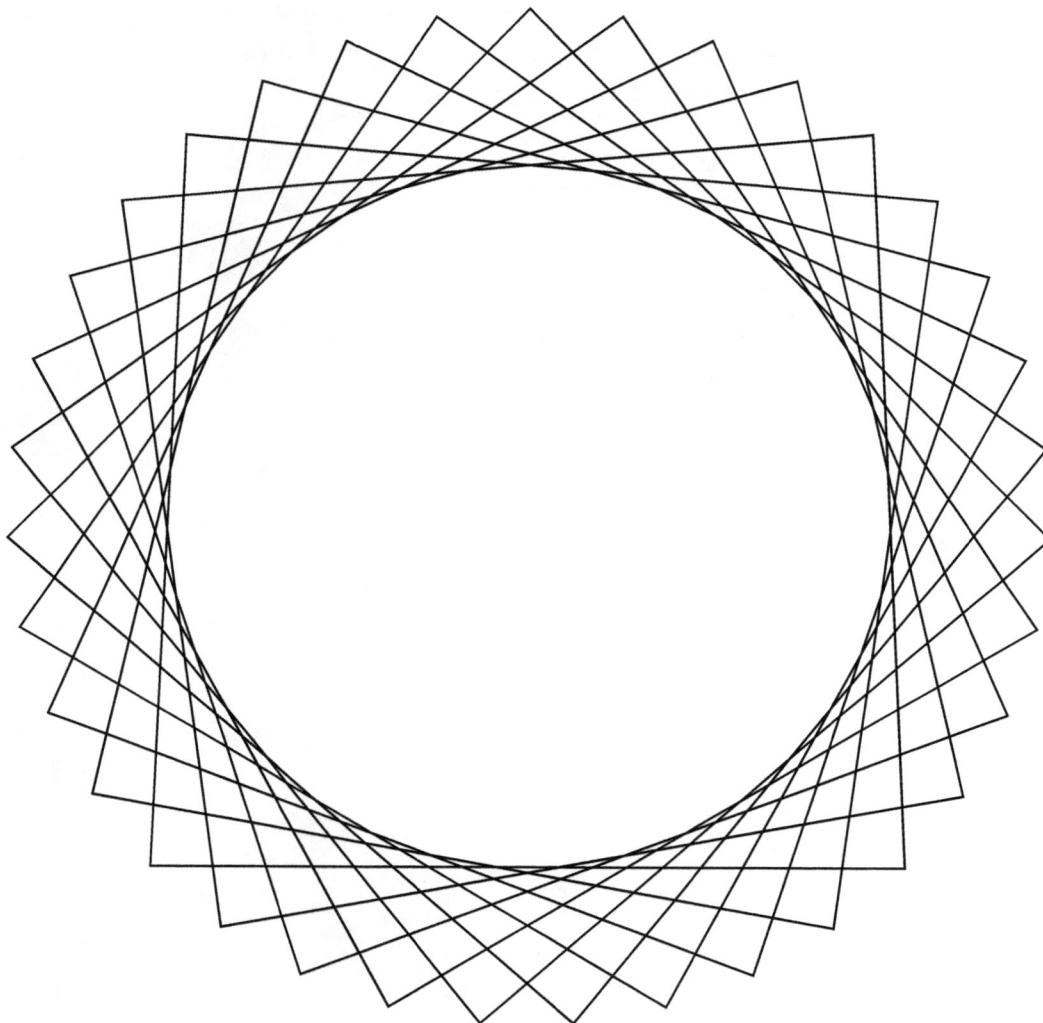

Figure 5.5   AAC [35: 0 9]

5.4.2 Point and angle data

Listing 5.3 shows the point and angle data for the AAC.

Listing 5.3

```
pt [ 0] = (10.000 10.000); a[0] a[1] =    0   9
pt [ 1] = (20.000 10.000); a[0] a[1] =    9   9
pt [ 2] = (19.551 19.990); a[0] a[1] =   18   9
pt [ 3] = ( 9.592 19.094); a[0] a[1] =   27   9
pt [ 4] = (10.934  9.184); a[0] a[1] =    1   9
pt [ 5] = (20.773 10.970); a[0] a[1] =   10   9
pt [ 6] = (18.548 20.719); a[0] a[1] =   19   9
pt [ 7] = ( 8.908 18.059); a[0] a[1] =   28   9
pt [ 8] = (11.999  8.548); a[0] a[1] =    2   9
pt [ 9] = (21.361 12.062); a[0] a[1] =   11   9
pt [10] = (17.431 21.257); a[0] a[1] =   20   9
pt [11] = ( 8.421 16.918); a[0] a[1] =   29   9
pt [12] = (13.160  8.112); a[0] a[1] =    3   9
pt [13] = (21.744 13.241); a[0] a[1] =   12   9
pt [14] = (16.235 21.587); a[0] a[1] =   21   9
pt [15] = ( 8.145 15.709); a[0] a[1] =   30   9
pt [16] = (14.380  7.891); a[0] a[1] =    4   9
pt [17] = (21.911 14.470); a[0] a[1] =   13   9
pt [18] = (15.000 21.698); a[0] a[1] =   22   9
pt [19] = ( 8.089 14.470); a[0] a[1] =   31   9
pt [20] = (15.620  7.891); a[0] a[1] =    5   9
pt [21] = (21.855 15.709); a[0] a[1] =   14   9
pt [22] = (13.765 21.587); a[0] a[1] =   23   9
pt [23] = ( 8.256 13.241); a[0] a[1] =   32   9
pt [24] = (16.840  8.112); a[0] a[1] =    6   9
pt [25] = (21.579 16.918); a[0] a[1] =   15   9
pt [26] = (12.569 21.257); a[0] a[1] =   24   9
pt [27] = ( 8.639 12.062); a[0] a[1] =   33   9
pt [28] = (18.001  8.548); a[0] a[1] =    7   9
pt [29] = (21.092 18.059); a[0] a[1] =   16   9
pt [30] = (11.452 20.719); a[0] a[1] =   25   9
```

```
pt[31] = ( 9.227 10.970); a[0] a[1] =   34    9
pt[32] = (19.066  9.184); a[0] a[1] =    8    9
pt[33] = (20.408 19.094); a[0] a[1] =   17    9
pt[34] = (10.449 19.990); a[0] a[1] =   26    9

pt[35] = (10.000 10.000); a[0] a[1] =    0    9
```

5.4.3 Trace

The PRTS algorithm predicts the number of rotational symmetries of AAC $[A : a_0\ a_1] = [35 : 0\ 9]$ as follows:

— initialize —

(1) set the number of iterations per a_1 rollover

$nipr_1 = 1$

— $n = 0$ —

(2) find the change in a_0 after $nipr_1$ iterations

$a_0' = \text{EFDM}(A, N - n, < a_0\ a_1 >, nipr_1) = \text{EFDM}(35, 1, < 0\ 9 >, 1) = 9$

$da_0 = (a_0' - a_0)\ \bmod A = (9 - 0)\ \bmod 35 = 9$

(3) find the number of a_1 rollovers per $< a_0\ a_1 >$ rollover

$nr_0 = \text{LCM}(A, da_0)/da_0 = \text{LCM}(35, 9)/9 = 35$

(4) find the number of iterations per $< a_0\ a_1 >$ rollover

$nipr_0 = nipr_1 * nr_0 = 1 * 35 = 35$

— conclude —

(5) return the predicted number of AAC rotational symmetries

$prts = nr_0 = 35$

5.4.4 Discussion

Step 3: for this AAC, $\text{LCM}(A, da_0) = A * da_0 = 35 * 9 = 315$. This is in contrast to the AAC in example 1, where $\text{LCM}(A, da_0) = A = 35$, and the AAC in example 2, where $\text{LCM}(A, da_0) = 70$, which is greater than A but less than $A * da_0$. An examination of Figure 5.5 on page 96 and Listing 5.3 provides an intuitive basis for the origin of the rotational symmetries of order 1 AACs where $\text{LCM}(A, da_0) = A * da_0$.

5.5 PRTS example 4

5.5.1 Graph

Figure 5.6 shows an order 2 AAC with five rotational symmetries.

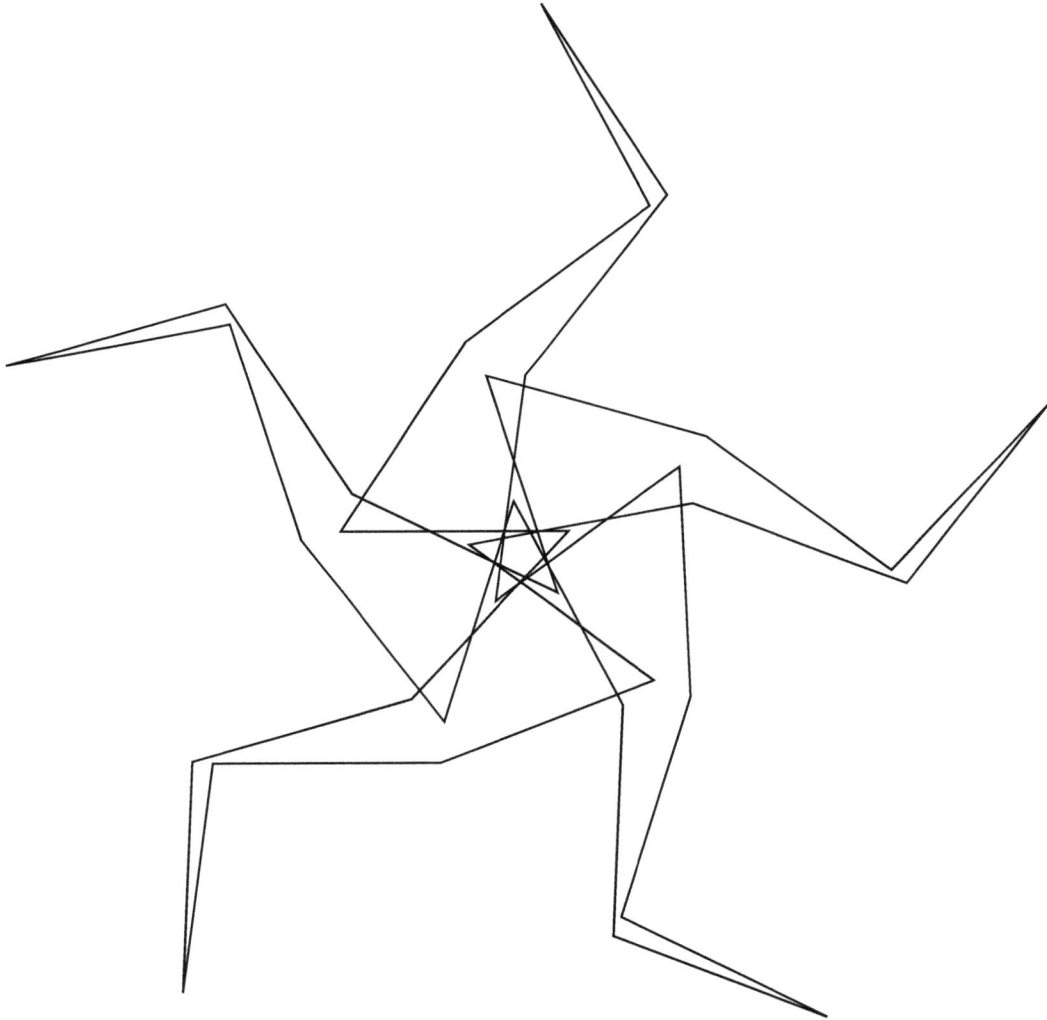

Figure 5.6   AAC [35: 0 22 10]

5.5.2   Point and angle data

Listing 5.4 shows the point and angle data for the AAC.

Listing 5.4

```
pt[ 0] = ( 10.000  10.000); a[0] ... a[2] =   0  22  10
pt[ 1] = ( 20.000  10.000); a[0] ... a[2] =  22  32  10
pt[ 2] = ( 13.089   2.772); a[0] ... a[2] =  19   7  10
pt[ 3] = (  3.450   0.112); a[0] ... a[2] =  26  17  10
pt[ 4] = (  3.001  -9.878); a[0] ... a[2] =   8  27  10
pt[ 5] = (  4.343   0.031); a[0] ... a[2] =   0   2  10
pt[ 6] = ( 14.343   0.031); a[0] ... a[2] =   2  12  10
pt[ 7] = ( 23.706   3.545); a[0] ... a[2] =  14  22  10
pt[ 8] = ( 15.616   9.423); a[0] ... a[2] =   1  32  10
pt[ 9] = ( 25.455  11.208); a[0] ... a[2] =  33   7  10
pt[10] = ( 34.817   7.695); a[0] ... a[2] =   5  17  10
pt[11] = ( 41.052  15.513); a[0] ... a[2] =  22  27  10
pt[12] = ( 34.142   8.285); a[0] ... a[2] =  14   2  10
pt[13] = ( 26.051  14.163); a[0] ... a[2] =  16  12  10
pt[14] = ( 16.412  16.823); a[0] ... a[2] =  28  22  10
pt[15] = ( 19.502   7.313); a[0] ... a[2] =  15  32  10
pt[16] = ( 10.492  11.652); a[0] ... a[2] =  12   7  10
pt[17] = (  4.983  19.997); a[0] ... a[2] =  19  17  10
pt[18] = ( -4.656  17.337); a[0] ... a[2] =   1  27  10
pt[19] = (  5.183  19.122); a[0] ... a[2] =  28   2  10
pt[20] = (  8.273   9.612); a[0] ... a[2] =  30  12  10
pt[21] = ( 14.508   1.794); a[0] ... a[2] =   7  22  10
pt[22] = ( 17.598  11.304); a[0] ... a[2] =  29  32  10
pt[23] = ( 22.337   2.498); a[0] ... a[2] =  26   7  10
pt[24] = ( 21.888  -7.492); a[0] ... a[2] =  33  17  10
pt[25] = ( 31.251 -11.005); a[0] ... a[2] =  15  27  10
pt[26] = ( 22.241  -6.667); a[0] ... a[2] =   7   2  10
pt[27] = ( 25.331   2.844); a[0] ... a[2] =   9  12  10
pt[28] = ( 24.882  12.834); a[0] ... a[2] =  21  22  10
pt[29] = ( 16.792   6.956); a[0] ... a[2] =   8  32  10
pt[30] = ( 18.135  16.865); a[0] ... a[2] =   5   7  10
```

```
pt[31] = ( 24.369   24.684); a[0] ... a[2] =   12   17   10
pt[32] = ( 18.860   33.030); a[0] ... a[2] =   29   27   10
pt[33] = ( 23.599   24.224); a[0] ... a[2] =   21    2   10
pt[34] = ( 15.509   18.346); a[0] ... a[2] =   23   12   10

pt[35] = ( 10.000   10.000); a[0] ... a[2] =    0   22   10
```

5.5.3   Trace

The PRTS algorithm predicts the number of rotational symmetries of AAC $[A : a_0\ a_1\ a_2] = [35 : 0\ 22\ 10]$ as follows:

— initialize —

(1) set the number of iterations per a_2 rollover

$nipr_2 = 1$

— $n = 1$ —

(2) find the change in a_1 after $nipr_2$ iterations

$a_1' = \text{EFDM}(A, N - n, < a_1\ a_2 >, nipr_2) = \text{EFDM}(35, 1, < 22\ 10 >, 1) = 32$

$da_1 = (a_1' - a_1) \mod A = (32 - 22) \mod 35 = 10$

(3) find the number of a_2 rollovers per $< a_1\ a_2 >$ rollover

$nr_1 = \text{LCM}(A, da_1)/da_1 = \text{LCM}(35, 10)/10 = 7$

(4) find the number of iterations per $< a_1\ a_2 >$ rollover

$nipr_1 = nipr_2 * nr_1 = 1 * 7 = 7$

— $n = 0$ —

(5) find the change in a_0 after $nipr_1$ iterations

$a_0' = \text{EFDM}(A, N - n, < a_0\ a_1\ a_2 >, nipr_1) = \text{EFDM}(35, 2, < 0\ 22\ 10 >, 7) = 14$

$da_0 = (a_0' - a_0) \mod A = (14 - 0) \mod 35 = 14$

(6) find the number of $< a_1\ a_2 >$ rollovers per $< a_0\ a_1\ a_2 >$ rollover

$nr_0 = \text{LCM}(A, da_0)/da_0 = \text{LCM}(35, 14)/14 = 5$

(7) find the number of iterations per $< a_0\ a_1\ a_2 >$ rollover

$nipr_0 = nipr_1 * nr_0 = 7 * 5 = 35$

— conclude —

(8) return the predicted number of AAC rotational symmetries

$prts = nr_0 = 5$

5.5.4 Discussion

Predicting the rotational symmetry of an AAC with order > 1 involves recursively applying AAC order 1 rotational symmetry prediction to higher difference orders. This example shows how this technique can be applied to an order 2 AAC.

Step 1: angle $a_2 = 10$ rolls over every iteration, as can be seen in Listing 5.4.

Step 2: the trace shows the general method of finding da_1, the change in a_1 after $nipr_2 = 1$ iteration, using the EFDM function. However, this method is unnecessarily complex for this difference order of the AAC. Listing 5.4 shows that the change in a_1 after one iteration is simply a_2, which is 10.

Step 3: with $A = 35$ slices per circle and a_1 increasing by $da_1 = 10$ slices each time a_2 rolls over, $< a_1\ a_2 >$ will roll over after $\text{LCM}(A, da_1)/da_1 = \text{LCM}(35, 10)/10 = \frac{70}{10} = 7$ a_2 rollovers.

Step 4: with $nipr_2 = 1$ iteration per a_2 rollover and $nr_1 = 7$ a_2 rollovers per $< a_1\ a_2 >$ rollover, there will be $nipr_1 = 7$ iterations per $< a_1\ a_2 >$ rollover. This can be verified in Listing 5.4, where $< a_1\ a_2 >=< 22\ 10 >$ indeed rolls over every 7 iterations.

Step 5: given that $< a_1\ a_2 >$ rolls over every 7 iterations, what is the change in a_0 each time $< a_1\ a_2 >$ rolls over? Inspection of Listing 5.4 yields an answer of 14. Fortunately, there is no need to actually iterate the angles $< a_0\ a_1\ a_2 >$ 7 times to determine a_0'. Instead, the PRTS function uses the EFDM function to calculate this value.

Step 6: with $A = 35$ slices per circle and a_0 increasing by $da_0 = 14$ slices each time $< a_1\ a_2 >$ rolls over, $< a_0\ a_1\ a_2 >$ will roll over after $\text{LCM}(A, da_0)/da_0 = \text{LCM}(35, 14)/14 = \frac{70}{14} = 5$ $< a_1\ a_2 >$ rollovers.

Step 7: with $nipr_1 = 7$ iterations per $< a_1\ a_2 >$ rollover and $nr_0 = 5 < a_1\ a_2 >$ rollovers per $< a_0\ a_1\ a_2 >$ rollover, there will be $nipr_0 = 35$ iterations per $< a_0\ a_1\ a_2 >$ rollover. This can be verified in Listing 5.4, where $< a_0\ a_1\ a_2 >=< 0\ 22\ 10 >$ indeed rolls over after 35 iterations.

Step 8: the order 2 AAC's predicted number of rotational symmetries is the predicted number of $< a_1\ a_2 >$ rollovers per $< a_0\ a_1\ a_2 >$ rollover, which is 5.

5.6   PRTS example 5

5.6.1   Graph

Figure 5.7 shows an order 3 AAC with six rotational symmetries.

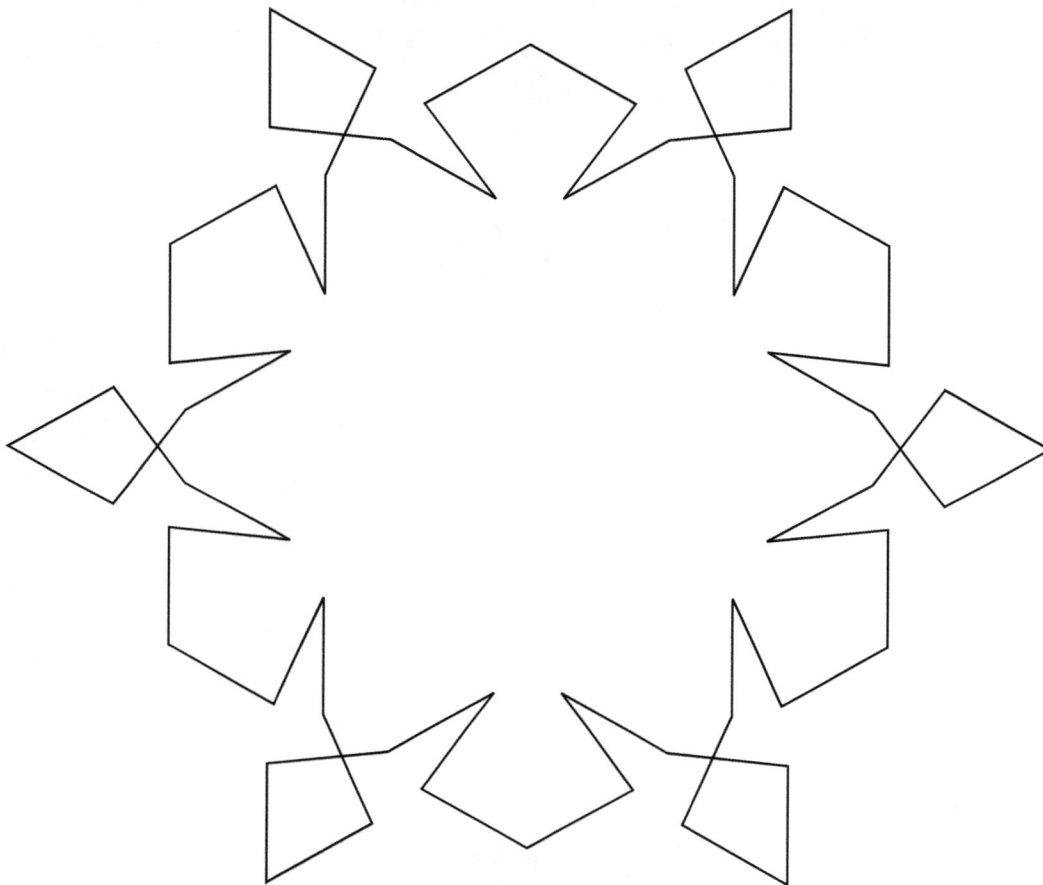

Figure 5.7   AAC [60: 9 26 30 48]

5.6.2   Point and angle data

Listing 5.5 shows the point and angle data for the AAC.

Listing 5.5

```
pt[ 0] = ( 10.000   10.000); a[0] ... a[3] =    9  26  30  48
pt[ 1] = ( 15.878   18.090); a[0] ... a[3] =   35  56  18  48
pt[ 2] = (  7.218   13.090); a[0] ... a[3] =   31  14   6  48
pt[ 3] = ( -2.728   12.045); a[0] ... a[3] =   45  20  54  48
pt[ 4] = ( -2.728    2.045); a[0] ... a[3] =    5  14  42  48
pt[ 5] = (  5.933    7.045); a[0] ... a[3] =   19  56  30  48
pt[ 6] = (  1.865   16.180); a[0] ... a[3] =   15  26  18  48
pt[ 7] = (  1.865   26.180); a[0] ... a[3] =   41  44   6  48
pt[ 8] = ( -2.202   17.045); a[0] ... a[3] =   25  50  54  48
pt[ 9] = (-10.862   22.045); a[0] ... a[3] =   15  44  42  48
pt[10] = (-10.862   32.045); a[0] ... a[3] =   59  26  30  48
pt[11] = ( -0.917   31.000); a[0] ... a[3] =   25  56  18  48
pt[12] = ( -9.577   36.000); a[0] ... a[3] =   21  14   6  48
pt[13] = (-15.455   44.090); a[0] ... a[3] =   35  20  54  48
pt[14] = (-24.115   39.090); a[0] ... a[3] =   55  14  42  48
pt[15] = (-15.455   34.090); a[0] ... a[3] =    9  56  30  48
pt[16] = ( -9.577   42.180); a[0] ... a[3] =    5  26  18  48
pt[17] = ( -0.917   47.180); a[0] ... a[3] =   31  44   6  48
pt[18] = (-10.862   46.135); a[0] ... a[3] =   15  50  54  48
pt[19] = (-10.862   56.135); a[0] ... a[3] =    5  44  42  48
pt[20] = ( -2.202   61.135); a[0] ... a[3] =   49  26  30  48
pt[21] = (  1.865   51.999); a[0] ... a[3] =   15  56  18  48
pt[22] = (  1.865   61.999); a[0] ... a[3] =   11  14   6  48
pt[23] = (  5.933   71.135); a[0] ... a[3] =   25  20  54  48
pt[24] = ( -2.728   76.135); a[0] ... a[3] =   45  14  42  48
pt[25] = ( -2.728   66.135); a[0] ... a[3] =   59  56  30  48
pt[26] = (  7.218   65.089); a[0] ... a[3] =   55  26  18  48
pt[27] = ( 15.878   60.089); a[0] ... a[3] =   21  44   6  48
pt[28] = ( 10.000   68.180); a[0] ... a[3] =    5  50  54  48
pt[29] = ( 18.660   73.180); a[0] ... a[3] =   55  44  42  48
pt[30] = ( 27.321   68.180); a[0] ... a[3] =   39  26  30  48
```

```
pt[31] = ( 21.443  60.089); a[0] ... a[3] =   5  56  18  48
pt[32] = ( 30.103  65.089); a[0] ... a[3] =   1  14   6  48
pt[33] = ( 40.048  66.135); a[0] ... a[3] =  15  20  54  48
pt[34] = ( 40.048  76.135); a[0] ... a[3] =  35  14  42  48
pt[35] = ( 31.388  71.135); a[0] ... a[3] =  49  56  30  48
pt[36] = ( 35.455  61.999); a[0] ... a[3] =  45  26  18  48
pt[37] = ( 35.455  51.999); a[0] ... a[3] =  11  44   6  48
pt[38] = ( 39.523  61.135); a[0] ... a[3] =  55  50  54  48
pt[39] = ( 48.183  56.135); a[0] ... a[3] =  45  44  42  48
pt[40] = ( 48.183  46.135); a[0] ... a[3] =  29  26  30  48
pt[41] = ( 38.238  47.180); a[0] ... a[3] =  55  56  18  48
pt[42] = ( 46.898  42.180); a[0] ... a[3] =  51  14   6  48
pt[43] = ( 52.776  34.090); a[0] ... a[3] =   5  20  54  48
pt[44] = ( 61.436  39.090); a[0] ... a[3] =  25  14  42  48
pt[45] = ( 52.776  44.090); a[0] ... a[3] =  39  56  30  48
pt[46] = ( 46.898  36.000); a[0] ... a[3] =  35  26  18  48
pt[47] = ( 38.238  31.000); a[0] ... a[3] =   1  44   6  48
pt[48] = ( 48.183  32.045); a[0] ... a[3] =  45  50  54  48
pt[49] = ( 48.183  22.045); a[0] ... a[3] =  35  44  42  48
pt[50] = ( 39.523  17.045); a[0] ... a[3] =  19  26  30  48
pt[51] = ( 35.455  26.180); a[0] ... a[3] =  45  56  18  48
pt[52] = ( 35.455  16.180); a[0] ... a[3] =  41  14   6  48
pt[53] = ( 31.388   7.045); a[0] ... a[3] =  55  20  54  48
pt[54] = ( 40.048   2.045); a[0] ... a[3] =  15  14  42  48
pt[55] = ( 40.048  12.045); a[0] ... a[3] =  29  56  30  48
pt[56] = ( 30.103  13.090); a[0] ... a[3] =  25  26  18  48
pt[57] = ( 21.443  18.090); a[0] ... a[3] =  51  44   6  48
pt[58] = ( 27.321  10.000); a[0] ... a[3] =  35  50  54  48
pt[59] = ( 18.660   5.000); a[0] ... a[3] =  25  44  42  48

pt[60] = ( 10.000  10.000); a[0] ... a[3] =   9  26  30  48
```

5.6.3 Trace

The PRTS algorithm predicts the number of rotational symmetries of
AAC $[A : a_0\ a_1\ a_2\ a_3] = [60 : 9\ 26\ 30\ 48]$ as follows:

— initialize —
(1) set the number of iterations per a_3 rollover
   $nipr_3 = 1$

— $n = 2$ —
(2) find the change in a_2 after $nipr_3$ iterations
   $a_2' = \text{EFDM}(A, N - n, < a_2\ a_3 >, nipr_3) = \text{EFDM}(60, 1, < 30\ 48 >, 1) = 18$
   $da_2 = (a_2' - a_2)\ \text{mod}\ A = (18 - 30)\ \text{mod}\ 60 = 48$
(3) find the number of a_3 rollovers per $< a_2\ a_3 >$ rollover
   $nr_2 = \text{LCM}(A, da_2)/da_2 = \text{LCM}(60, 48)/48 = 5$
(4) find the number of iterations per $< a_2\ a_3 >$ rollover
   $nipr_2 = nipr_3 * nr_2 = 1 * 5 = 5$

— $n = 1$ —
(5) find the change in a_1 after $nipr_2$ iterations
   $a_1' = \text{EFDM}(A, N - n, < a_1\ a_2\ a_3 >, nipr_2) = \text{EFDM}(60, 2, < 26\ 30\ 48 >, 5) = 56$
   $da_1 = (a_1' - a_1)\ \text{mod}\ A = (56 - 26)\ \text{mod}\ 60 = 30$
(6) find the number of $< a_2\ a_3 >$ rollovers per $< a_1\ a_2\ a_3 >$ rollover
   $nr_1 = \text{LCM}(A, da_1)/da_1 = \text{LCM}(60, 30)/30 = 2$
(7) find the number of iterations per $< a_1\ a_2\ a_3 >$ rollover
   $nipr_1 = nipr_2 * nr_1 = 5 * 2 = 10$

— $n = 0$ —
(8) find the change in a_0 after $nipr_1$ iterations
   $a_0' = \text{EFDM}(A, N - n, < a_0\ a_1\ a_2\ a_3 >, nipr_1) = \text{EFDM}(60, 3, < 9\ 26\ 30\ 48 >, 10) = 59$
   $da_0 = (a_0' - a_0)\ \text{mod}\ A = (59 - 9)\ \text{mod}\ 60 = 50$
(9) find the number of $< a_1\ a_2\ a_3 >$ rollovers per $< a_0\ a_1\ a_2\ a_3 >$ rollover
   $nr_0 = \text{LCM}(A, da_0)/da_0 = \text{LCM}(60, 50)/50 = 6$
(10) find the number of iterations per $< a_0\ a_1\ a_2\ a_3 >$ rollover
   $nipr_0 = nipr_1 * nr_0 = 10 * 6 = 60$

— conclude —

(11) return the predicted number of AAC rotational symmetries

$$j = nr_0 = 6$$

5.6.4 Discussion

Predicting the rotational symmetry of an AAC with order > 1 involves recursively applying AAC order 1 rotational symmetry prediction to higher difference orders. This example shows how this technique can be applied to an order 3 AAC.

Step 1: angle $a_3 = 48$ rolls over every iteration, as can be seen in Listing 5.5.

Step 2: the trace shows the general method of finding da_2, the change in a_2 after $nipr_3 = 1$ iteration, using the EFDM function. However, this method is unnecessarily complex for this difference order of the AAC. Listing 5.5 shows that the change in a_2 after one iteration is simply a_3, which is 48.

Step 3: with $A = 60$ slices per circle and a_2 increasing by $da_2 = 48$ slices each time a_3 rolls over, $< a_2\ a_3 >$ will roll over after $nr_2 = \text{LCM}(A, da_2)/da_2 = \text{LCM}(60, 48)/48 = \frac{240}{48} = 5$ a_3 rollovers.

Step 4: with $nipr_3 = 1$ iteration per a_3 rollover and $nr_2 = 5$ a_3 rollovers per $< a_2\ a_3 >$ rollover, there will be $nipr_2 = 5$ iterations per $< a_2\ a_3 >$ rollover. This can be verified in Listing 5.5, where $< a_2\ a_3 > = < 30\ 48 >$ indeed rolls over every 5 iterations.

Step 5: given that $< a_2\ a_3 >$ rolls over every 5 iterations, what is the change in a_1 each time $< a_2\ a_3 >$ rolls over? The PRTS function uses the EFDM function to calculate that a_1' is 56, and subsequently concludes that da_1 is 30. This can be verified in Listing 5.5.

Step 6: With $A = 60$ slices per circle and a_1 increasing by $da_1 = 30$ slices each time $< a_2\ a_3 >$ rolls over, $< a_1\ a_2\ a_3 >$ will roll over after $nr_1 = \text{LCM}(A, da_1)/da_1 = \text{LCM}(60, 30)/30 = \frac{60}{30} = 2$ $< a_2\ a_3 >$ rollovers.

Step 7: with $nipr_2 = 5$ iterations per $< a_2\ a_3 >$ rollover and $nr_1 = 2 < a_2\ a_3 >$ rollovers per $< a_1\ a_2\ a_3 >$ rollover, there will be $nipr_1 = 10$ iterations per $< a_1\ a_2\ a_3 >$ rollover. This can be verified in Listing 5.5, where $< a_1\ a_2\ a_3 > = < 26\ 30\ 48 >$ indeed rolls over every 10 iterations.

Step 8: given that $< a_1\ a_2\ a_3 >$ rolls over every 10 iterations, what is the change in a_0 each time $< a_1\ a_2\ a_3 >$ rolls over? The PRTS function uses the EFDM function to calculate that a'_0 is 59, and subsequently concludes that da_0 is 50. This can be verified in Listing 5.5.

Step 9: with $A = 60$ slices per circle and a_0 increasing by $da_0 = 50$ slices each time $< a_1\ a_2\ a_3 >$ rolls over, $< a_0\ a_1\ a_2\ a_3 >$ will roll over after $nr_0 = \text{LCM}(A, da_0)/da_0 = \text{LCM}(60, 50)/50 = \frac{300}{50} = 6 < a_1\ a_2\ a_3 >$ rollovers.

Step 10: with $nipr_1 = 10$ iterations per $< a_1\ a_2\ a_3 >$ rollover and $nr_0 = 6 < a_1\ a_2\ a_3 >$ rollovers per $< a_0\ a_1\ a_2\ a_3 >$ rollover, there will be $nipr_0 = 60$ iterations per $< a_0\ a_1\ a_2\ a_3 >$ rollover. This can be verified in Listing 5.5, where $< a_0\ a_1\ a_2\ a_3 > = < 9\ 26\ 30\ 48 >$ indeed rolls over after 60 iterations.

Step 11: the order 3 AAC's predicted number of rotational symmetries is the predicted number of $< a_1\ a_2\ a_3 >$ rollovers per $< a_0\ a_1\ a_2\ a_3 >$ rollover, which is 6.

5.7 PRTS example 6

5.7.1 Graph

Figure 5.8 shows an order 3 AAC with 7 rotational symmetries.

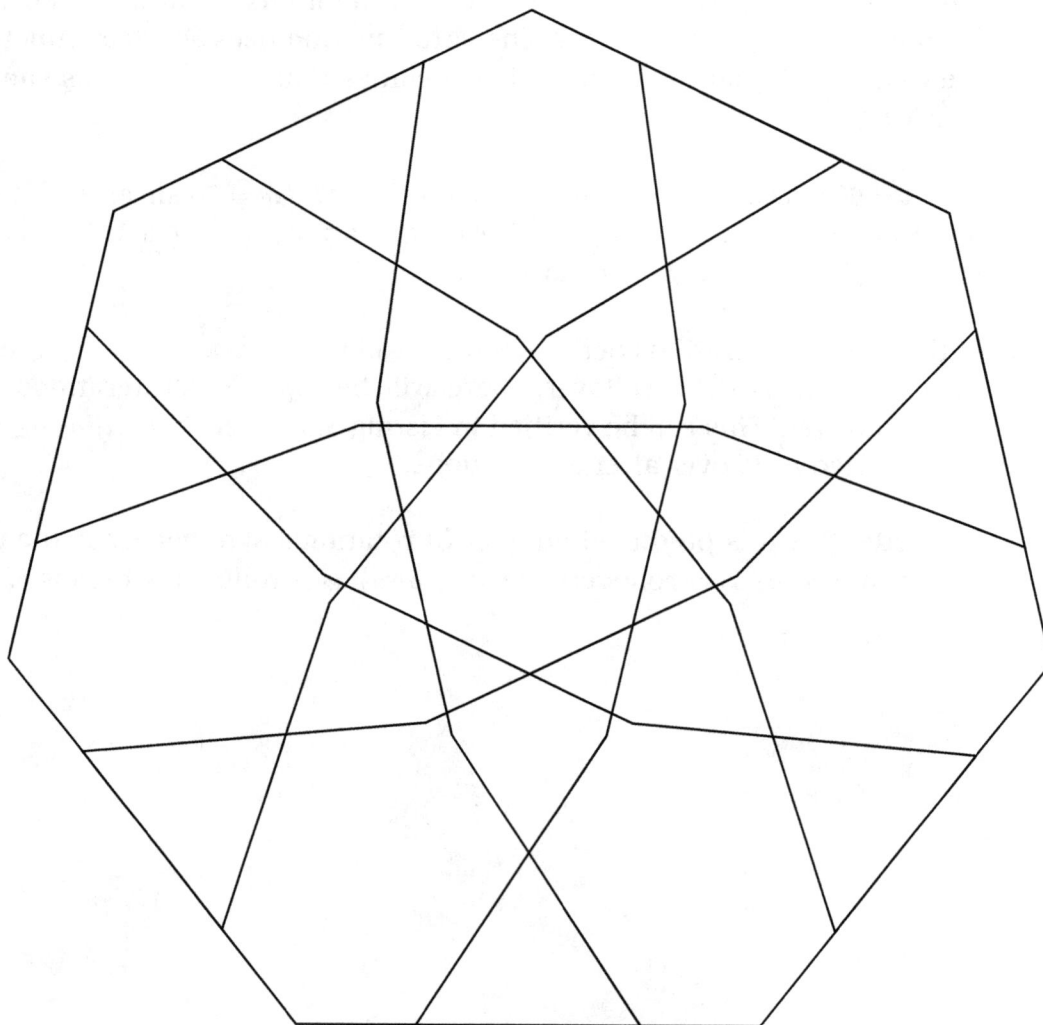

Figure 5.8 AAC [35: 0 12 21 14]

5.7.2 Point and angle data

Listing 5.6 shows the point and angle data for the AAC.

Listing 5.6

```
pt[ 0] = ( 10.000  10.000); a[0] ... a[3] =   0  12  21  14
pt[ 1] = ( 20.000  10.000); a[0] ... a[3] =  12  33   0  14
pt[ 2] = ( 14.491  18.346); a[0] ... a[3] =  10  33  14  14
pt[ 3] = ( 12.266  28.095); a[0] ... a[3] =   8  12  28  14
pt[ 4] = ( 13.608  38.005); a[0] ... a[3] =  20   5   7  14
pt[ 5] = (  4.598  33.666); a[0] ... a[3] =  25  12  21  14
pt[ 6] = (  2.373  23.916); a[0] ... a[3] =   2  33   0  14
pt[ 7] = ( 11.736  27.430); a[0] ... a[3] =   0  33  14  14
pt[ 8] = ( 21.736  27.430); a[0] ... a[3] =  33  12  28  14
pt[ 9] = ( 31.098  23.916); a[0] ... a[3] =  10   5   7  14
pt[10] = ( 28.873  33.666); a[0] ... a[3] =  15  12  21  14
pt[11] = ( 19.863  38.005); a[0] ... a[3] =  27  33   0  14
pt[12] = ( 21.205  28.095); a[0] ... a[3] =  25  33  14  14
pt[13] = ( 18.980  18.346); a[0] ... a[3] =  23  12  28  14
pt[14] = ( 13.471  10.000); a[0] ... a[3] =   0   5   7  14
pt[15] = ( 23.471  10.000); a[0] ... a[3] =   5  12  21  14
pt[16] = ( 29.706  17.818); a[0] ... a[3] =  17  33   0  14
pt[17] = ( 19.746  18.715); a[0] ... a[3] =  15  33  14  14
pt[18] = ( 10.737  23.054); a[0] ... a[3] =  13  12  28  14
pt[19] = (  3.826  30.281); a[0] ... a[3] =  25   5   7  14
pt[20] = (  1.601  20.532); a[0] ... a[3] =  30  12  21  14
pt[21] = (  7.836  12.714); a[0] ... a[3] =   7  33   0  14
pt[22] = ( 10.926  22.224); a[0] ... a[3] =   5  33  14  14
pt[23] = ( 17.161  30.043); a[0] ... a[3] =   3  12  28  14
pt[24] = ( 25.745  35.172); a[0] ... a[3] =  15   5   7  14
pt[25] = ( 16.736  39.511); a[0] ... a[3] =  20  12  21  14
pt[26] = (  7.726  35.172); a[0] ... a[3] =  32  33   0  14
pt[27] = ( 16.310  30.043); a[0] ... a[3] =  30  33  14  14
pt[28] = ( 22.545  22.224); a[0] ... a[3] =  28  12  28  14
pt[29] = ( 25.635  12.714); a[0] ... a[3] =   5   5   7  14
pt[30] = ( 31.870  20.532); a[0] ... a[3] =  10  12  21  14
```

```
pt[31] = ( 29.645  30.281); a[0] ... a[3] =  22  33   0  14
pt[32] = ( 22.735  23.054); a[0] ... a[3] =  20  33  14  14
pt[33] = ( 13.725  18.715); a[0] ... a[3] =  18  12  28  14
pt[34] = (  3.765  17.818); a[0] ... a[3] =  30   5   7  14

pt[35] = ( 10.000  10.000); a[0] ... a[3] =   0  12  21  14
```

5.7.3 Trace

The PRTS algorithm predicts the number of rotational symmetries of
AAC $[A : a_0 \ a_1 \ a_2 \ a_3] = [35 : 0 \ 12 \ 21 \ 14]$ as follows:

— initialize —

(1) set the number of iterations per a_3 rollover

$nipr_3 = 1$

— n = 2 —

(2) find the change in a_2 after $nipr_3$ iterations

$a_2' = \text{EFDM}(A, N - n, < a_2 \ a_3 >, nipr_3) = \text{EFDM}(35, 1, < 21 \ 14 >, 1) = 0$

$da_2 = (a_2' - a_2) \mod A = (0 - 21) \mod 35 = 14$

(3) find the number of a_3 rollovers per $< a_2 \ a_3 >$ rollover

$nr_2 = \text{LCM}(A, da_2)/da_2 = \text{LCM}(35, 14)/14 = 5$

(4) find the number of iterations per $< a_2 \ a_3 >$ rollover

$nipr_2 = nipr_3 * nr_2 = 1 * 5 = 5$

— n = 1 —

(5) find the change in a_1 after $nipr_2$ iterations

$a_1' = \text{EFDM}(A, N - n, < a_1 \ a_2 \ a_3 >, nipr_2) = \text{EFDM}(35, 2, < 12 \ 21 \ 14 >, 5) = 12$

$da_1 = (a_1' - a_1) \mod A = (12 - 12) \mod 35 = 0$

(6) find the number of $< a_2 \ a_3 >$ rollovers per $< a_1 \ a_2 \ a_3 >$ rollover

$nr_1 = 1$

(7) find the number of iterations per $< a_1 \ a_2 \ a_3 >$ rollover

$nipr_1 = nipr_2 * nr_1 = 5 * 1 = 5$

$$- n = 0 -$$

(8) find the change in a_0 after $nipr_1$ iterations

$a'_0 = \text{EFDM}(A, N - n, < a_0\ a_1\ a_2\ a_3 >, nipr_1) = \text{EFDM}(35, 3, < 0\ 12\ 21\ 14 >, 5) = 25$

$da_0 = (a'_0 - a_0)\ \bmod A = (25 - 0)\ \bmod 35 = 25$

(9) find the number of $< a_1\ a_2\ a_3 >$ rollovers per $< a_0\ a_1\ a_2\ a_3 >$ rollover

$nr_0 = \text{LCM}(A, da_0)/da_0 = \text{LCM}(35, 25)/25 = 7$

(10) find the number of iterations per $< a_0\ a_1\ a_2\ a_3 >$ rollover

$nipr_0 = nipr_1 * nr_0 = 5 * 7 = 35$

$$- \text{conclude} -$$

(11) return the predicted number of AAC rotational symmetries

$prts = nr_0 = 7$

5.7.4 Discussion

This example illustrates how the PRTS function handles the situation where da_n, the change in a_n after $nipr_{n+1}$ iterations, is 0. In this case, the difference order n is 1.

Step 5: given that $< a_2\ a_3 >$ rolls over every 5 iterations, what is the change in a_1 each time $< a_2\ a_3 >$ rolls over? The PRTS function uses the EFDM function to calculate that a'_1 is 12, and subsequently concludes that da_1 is 0. This can be verified in Listing 5.6.

Step 6: with a_1 increasing by $da_1 = 0$ slices each time $< a_2\ a_3 >$ rolls over, $< a_1\ a_2\ a_3 >$ will roll over each time $< a_2\ a_3 >$ rolls over, so nr_1 is set to 1. The standard method of calculating nr_1, $\text{LCM}(A, da_1)/da_1$, cannot be employed when da_1 is 0.

Step 7: with $nipr_2 = 5$ iterations per $< a_2\ a_3 >$ rollover and $nr_1 = 1 < a_2\ a_3 >$ rollover per $< a_1\ a_2\ a_3 >$ rollover, there will be $nipr_1 = 5$ iterations per $< a_1\ a_2\ a_3 >$ rollover. This can be verified in Listing 5.6, where $< a_1\ a_2\ a_3 >=< 12\ 21\ 14 >$ indeed rolls over every 5 iterations.

5.8 PRTS example 7

5.8.1 Graph

Figure 5.9 shows an order 3 AAC with only trivial rotational symmetry.

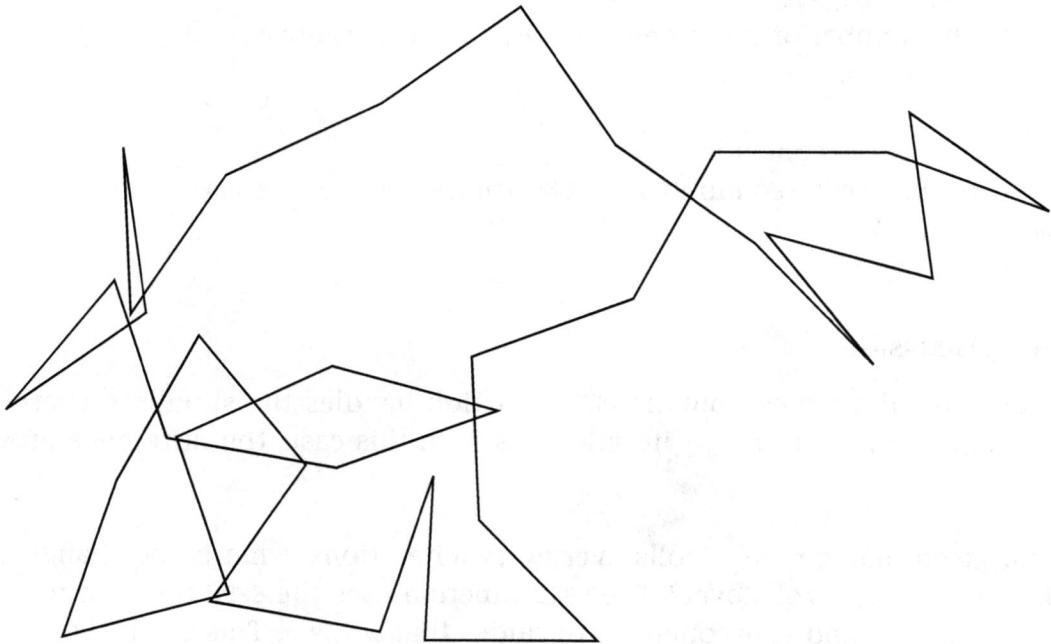

Figure 5.9 AAC [35: 0 13 18 14]

5.8.2 Point and angle data

Listing 5.7 shows the point and angle data for the AAC.

Listing 5.7

```
pt[ 0] = ( 10.000  10.000); a[0] ... a[3] =   0  13  18  14
pt[ 1] = ( 20.000  10.000); a[0] ... a[3] =  13  31  32  14
pt[ 2] = ( 13.089  17.228); a[0] ... a[3] =   9  28  11  14
pt[ 3] = ( 12.641  27.218); a[0] ... a[3] =   2   4  25  14
pt[ 4] = ( 22.003  30.732); a[0] ... a[3] =   6  29   4  14
pt[ 5] = ( 26.742  39.538); a[0] ... a[3] =   0  33  18  14
pt[ 6] = ( 36.742  39.538); a[0] ... a[3] =  33  16  32  14
pt[ 7] = ( 46.104  36.024); a[0] ... a[3] =  14  13  11  14
pt[ 8] = ( 38.014  41.902); a[0] ... a[3] =  27  24  25  14
pt[ 9] = ( 39.356  31.992); a[0] ... a[3] =  16  14   4  14
pt[10] = ( 29.717  34.653); a[0] ... a[3] =  30  18  18  14
pt[11] = ( 35.952  26.834); a[0] ... a[3] =  13   1  32  14
pt[12] = ( 29.041  34.062); a[0] ... a[3] =  14  33  11  14
pt[13] = ( 20.951  39.940); a[0] ... a[3] =  12   9  25  14
pt[14] = ( 15.442  48.286); a[0] ... a[3] =  21  34   4  14
pt[15] = (  7.352  42.408); a[0] ... a[3] =  20   3  18  14
pt[16] = ( -1.658  38.069); a[0] ... a[3] =  23  21  32  14
pt[17] = ( -7.167  29.723); a[0] ... a[3] =   9  18  11  14
pt[18] = ( -7.616  39.713); a[0] ... a[3] =  27  29  25  14
pt[19] = ( -6.273  29.804); a[0] ... a[3] =  21  19   4  14
pt[20] = (-14.364  23.926); a[0] ... a[3] =   5  23  18  14
pt[21] = ( -8.129  31.744); a[0] ... a[3] =  28   6  32  14
pt[22] = ( -5.038  22.234); a[0] ... a[3] =  34   3  11  14
pt[23] = (  4.801  20.448); a[0] ... a[3] =   2  14  25  14
pt[24] = ( 14.163  23.962); a[0] ... a[3] =  16   4   4  14
pt[25] = (  4.524  26.622); a[0] ... a[3] =  20   8  18  14
pt[26] = ( -4.486  22.283); a[0] ... a[3] =  28  26  32  14
pt[27] = ( -1.396  12.773); a[0] ... a[3] =  19  23  11  14
pt[28] = (-11.036  10.112); a[0] ... a[3] =   7  34  25  14
pt[29] = ( -7.945  19.623); a[0] ... a[3] =   6  24   4  14
pt[30] = ( -3.207  28.429); a[0] ... a[3] =  30  28  18  14
```

```
pt[31] = (  3.028  20.611); a[0] ... a[3] =  23  11  32  14
pt[32] = ( -2.481  12.265); a[0] ... a[3] =  34   8  11  14
pt[33] = (  7.358  10.479); a[0] ... a[3] =   7  19  25  14
pt[34] = ( 10.449  19.990); a[0] ... a[3] =  26   9   4  14

pt[35] = ( 10.000  10.000); a[0] ... a[3] =   0  13  18  14
```

5.8.3 Trace

The PRTS algorithm predicts the number of rotational symmetries of
AAC $[A : a_0\ a_1\ a_2\ a_3] = [35 : 0\ 13\ 18\ 14]$ as follows:

— initialize —

(1) set the number of iterations per a_3 rollover

$$nipr_3 = 1$$

— $n = 2$ —

(2) find the change in a_2 after $nipr_3$ iterations

$$a'_2 = \text{EFDM}(A, N - n, < a_2\ a_3 >, nipr_3) = \text{EFDM}(35, 1, < 18\ 14 >, 1) = 32$$
$$da_2 = (a'_2 - a_2)\ \ \text{mod}\ A = (18 - 32)\ \ \text{mod}\ 35 = 14$$

(3) find the number of a_3 rollovers per $< a_2\ a_3 >$ rollover

$$nr_2 = \text{LCM}(A, da_2)/da_2 = \text{LCM}(35, 14)/14 = 5$$

(4) find the number of iterations per $< a_2\ a_3 >$ rollover

$$nipr_2 = nipr_3 * nr_2 = 1 * 5 = 5$$

— $n = 1$ —

(5) find the change in a_1 after $nipr_2$ iterations

$$a'_1 = \text{EFDM}(A, N - n, < a_1\ a_2\ a_3 >, nipr_2) = \text{EFDM}(35, 2, < 13\ 18\ 14 >, 5) = 33$$
$$da_1 = (a'_1 - a_1)\ \ \text{mod}\ A = (33 - 13)\ \ \text{mod}\ 35 = 20$$

(6) find the number of $< a_2\ a_3 >$ rollovers per $< a_1\ a_2\ a_3 >$ rollover

$$nr_1 = \text{LCM}(A, da_1)/da_1 = \text{LCM}(35, 20)/20 = 7$$

(7) find the number of iterations per $< a_1\ a_2\ a_3 >$ rollover

$$nipr_1 = nipr_2 * nr_1 = 5 * 7 = 35$$

$$- n = 0 -$$

(8) find the change in a_0 after $nipr_1$ iterations

$a'_0 = \text{EFDM}(A, N - n, < a_0\ a_1\ a_2\ a_3 >, nipr_1) = \text{EFDM}(35, 3, < 0\ 13\ 18\ 14 >, 35) = 0$

$da_0 = (a'_0 - a_0)\ \mod A = (0 - 0)\ \mod 35 = 0$

(9) find the number of $< a_1\ a_2\ a_3 >$ rollovers per $< a_0\ a_1\ a_2\ a_3 >$ rollover

$nr_0 = 1$

(10) find the number of iterations per $< a_0\ a_1\ a_2\ a_3 >$ rollover

$nipr_0 = nipr_1 * nr_0 = 35 * 1 = 35$

$$- \text{conclude} -$$

(11) return the predicted number of AAC rotational symmetries

$prts = nr_0 = 1$

5.8.4 Discussion

Trace steps 8 – 11 show the origin of the AAC's nontrivial rotational asymmetry.

Step 8: the PRTS function calculates that each time $< a_1\ a_2\ a_3 > = < 13\ 18\ 14 >$ rolls over, the change in a_0, da_0, is 0. This can be verified by examining Listing 5.7.

Step 9: with a_0 increasing by $da_0 = 0$ slices each time $< a_1\ a_2\ a_3 >$ rolls over, $< a_0\ a_1\ a_2\ a_3 >$ will roll over each time $< a_1\ a_2\ a_3 >$ rolls over, so nr_0 is set to 1. The standard method of calculating nr_0, LCM$(A, da_0)/da_0$, cannot be employed when da_0 is 0.

Step 10: with $nipr_1 = 35$ iterations per $< a_1\ a_2\ a_3 >$ rollover and $nr_0 = 1 < a_1\ a_2\ a_3 >$ rollover per $< a_0\ a_1\ a_2\ a_3 >$ rollover, there will be $nipr_0 = 35$ iterations per $< a_0\ a_1\ a_2\ a_3 >$ rollover. This can be verified in Listing 5.7, where $< a_0\ a_1\ a_2\ a_3 > = < 0\ 13\ 18\ 14 >$ indeed rolls over after 35 iterations.

Step 11: the AAC's predicted number of rotational symmetries is the predicted number of $< a_1\ a_2\ a_3 >$ rollovers per $< a_0\ a_1\ a_2\ a_3 >$ rollover, which is 1.

6 Determining Accelerating Angle Curve Rotational Symmetry

This chapter provides a method of determining the number of rotational symmetries (DRTS) of a closed AAC given only the AAC's point data. The chapter begins with an introduction to DRTS, and then presents a DRTS algorithm. Finally, an example of DRTS algorithm operation is provided.

6.1 DRTS introduction

An AAC has an *angle a rotational symmetry* if rotating the entire AAC counterclockwise by a around the AAC's centroid results in a rotated AAC tracing the same path in the same direction in the xy plane as the original AAC. The points in an AAC's path are numbered in sequence from 0 to $np - 1$, where np is the number of points in the AAC. A rotational symmetry of an AAC therefore maps the AAC's points onto themselves. An *AAC map* consists of np individual point maps.

For example, the AAC in Figure 6.1 on the next page has six rotational symmetries. The AAC's points are numbered from 0 to 5. Let $\mathtt{rot}(p, a)$ be a function that rotates a point p by a degrees around the AAC's centroid. The six AAC maps, $aacm$, associated with the six rotational symmetries of the AAC are:

$aacm = 1.$ rotation angle $a = 60$ degrees:
$\quad \mathtt{rot}(pt[0], 60) = pt[1]$
$\quad \mathtt{rot}(pt[1], 60) = pt[2]$
$\quad \mathtt{rot}(pt[2], 60) = pt[3]$
$\quad \mathtt{rot}(pt[3], 60) = pt[4]$
$\quad \mathtt{rot}(pt[4], 60) = pt[5]$
$\quad \mathtt{rot}(pt[5], 60) = pt[0]$

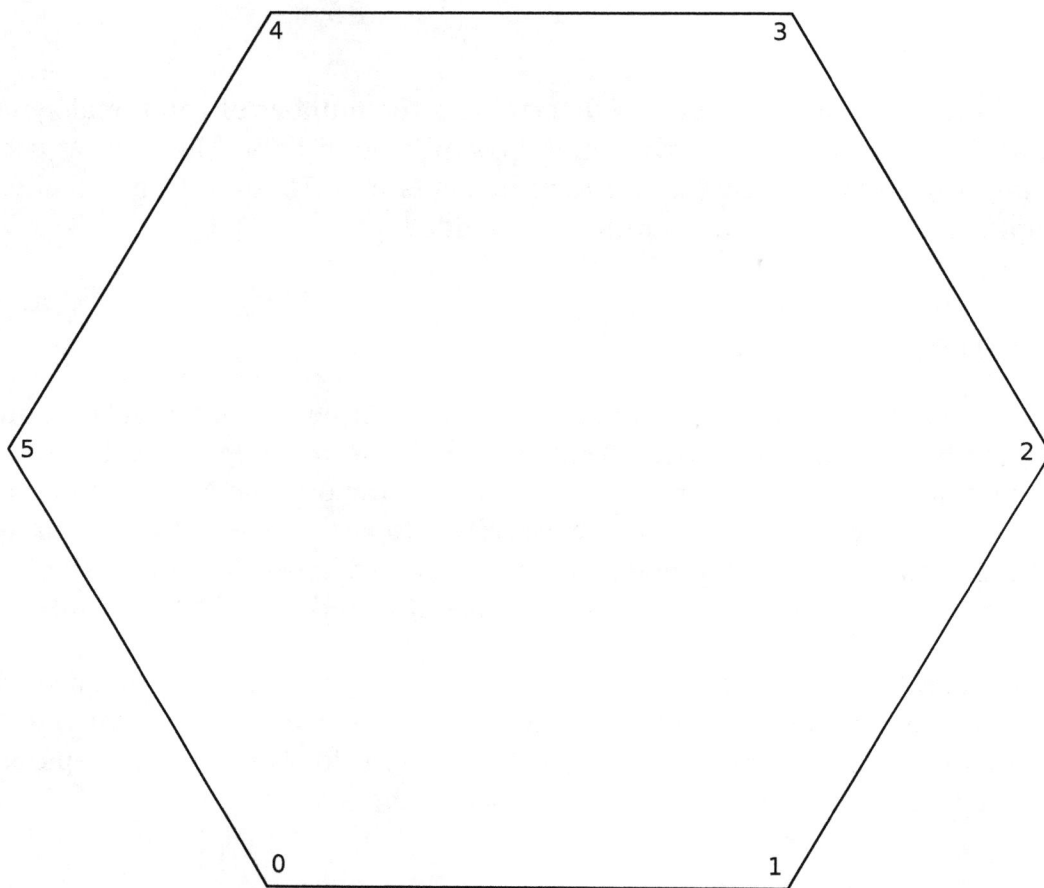

Figure 6.1 AAC [24: 0 4]

$aacm = 2$. rotation angle $a = 120$ degrees:

$\text{rot}(pt[0], 120) = pt[2]$
$\text{rot}(pt[1], 120) = pt[3]$
$\text{rot}(pt[2], 120) = pt[4]$
$\text{rot}(pt[3], 120) = pt[5]$
$\text{rot}(pt[4], 120) = pt[0]$
$\text{rot}(pt[5], 120) = pt[1]$

$aacm = 3$. rotation angle $a = 180$ degrees:

$\text{rot}(pt[0], 180) = pt[3]$
$\text{rot}(pt[1], 180) = pt[4]$
$\text{rot}(pt[2], 180) = pt[5]$
$\text{rot}(pt[3], 180) = pt[0]$
$\text{rot}(pt[4], 180) = pt[1]$
$\text{rot}(pt[5], 180) = pt[2]$

$aacm = 4$. rotation angle $a = 240$ degrees:

$\text{rot}(pt[0], 240) = pt[4]$
$\text{rot}(pt[1], 240) = pt[5]$
$\text{rot}(pt[2], 240) = pt[0]$
$\text{rot}(pt[3], 240) = pt[1]$
$\text{rot}(pt[4], 240) = pt[2]$
$\text{rot}(pt[5], 240) = pt[3]$

$aacm = 5$. rotation angle $a = 300$ degrees:

$\text{rot}(pt[0], 300) = pt[5]$
$\text{rot}(pt[1], 300) = pt[0]$
$\text{rot}(pt[2], 300) = pt[1]$
$\text{rot}(pt[3], 300) = pt[2]$
$\text{rot}(pt[4], 300) = pt[3]$
$\text{rot}(pt[5], 300) = pt[4]$

$aacm = 6$. rotation angle $a = 360$ degrees:

$\text{rot}(pt[0], 360) = pt[0]$

$\text{rot}(pt[1], 360) = pt[1]$

$\text{rot}(pt[2], 360) = pt[2]$

$\text{rot}(pt[3], 360) = pt[3]$

$\text{rot}(pt[4], 360) = pt[4]$

$\text{rot}(pt[5], 360) = pt[5]$

If a certain rotation angle a is a symmetric rotation of an AAC, then rotation angle $a' = c * a$, where c is a positive integer, is also a symmetric rotation of the AAC. This can be demonstrated for $c = 2$ by noting that if $\text{rot}(pt[0], 60) = pt[1]$ and $\text{rot}(pt[1], 60) = pt[2]$, then $\text{rot}(\text{rot}(pt[0], 60), 60) = pt[2]$. Similarly, if $\text{rot}(pt[1], 60) = pt[2]$ and $\text{rot}(pt[2], 60) = pt[3]$, then $\text{rot}(\text{rot}(pt[1], 60), 60) = pt[3]$. Similar relationships hold for the other points in the AAC. The complete AAC map is:

$\text{rot}(\text{rot}(pt[0], 60), 60) = pt[2]$

$\text{rot}(\text{rot}(pt[1], 60), 60) = pt[3]$

$\text{rot}(\text{rot}(pt[2], 60), 60) = pt[4]$

$\text{rot}(\text{rot}(pt[3], 60), 60) = pt[5]$

$\text{rot}(\text{rot}(pt[4], 60), 60) = pt[0]$

$\text{rot}(\text{rot}(pt[5], 60), 60) = pt[1]$

which can be reduced to the $aacm = 2$ symmetric rotation:

$\text{rot}(pt[0], 120) = pt[2]$

$\text{rot}(pt[1], 120) = pt[3]$

$\text{rot}(pt[2], 120) = pt[4]$

$\text{rot}(pt[3], 120) = pt[5]$

$\text{rot}(pt[4], 120) = pt[0]$

$\text{rot}(pt[5], 120) = pt[1]$

A similar analysis can be performed for $c = 3, 4, 5$, and 6.

When determining the number of rotational symmetries of an AAC, it is therefore only necessary to find the *smallest* angle a_{min} that is a rotational symmetry of the AAC. The total number of rotational symmetries of the AAC can then be calculated as $\frac{360}{a_{min}}$ if the rotation angle is measured in degrees, or $\frac{2\pi}{a_{min}}$ if the rotation angle is measured in radians.

6.2 DRTS algorithm

This section describes the DRTS algorithm main routine, DRTS, and two of its three subroutines, CNTRD and CRTS. Its third subroutine, LCM, is discussed in subsection 5.1.3 on page 89.

6.2.1 DTRS function

The *Determine Rotational Symmetry* function, DRTS, determines the number of rotational symmetries of a closed AAC. Pseudocode for the function is provided in Figure 6.2 on the next page.

The inputs to the function are the number of points in the AAC, np, and an array of points defining the AAC, $pt[0] \cdots pt[np-1]$. The line segment s start point and end point are $pt[s]$ and $pt[s+1]$, respectively, except for the last line segment in the AAC, where the end point is $pt[0]$.

The function initializes the working number of symmetric rotations found, $nsrtw$, to 1, and then scans all possible AAC maps for opportunities to increase $nsrtw$. When such an opportunity is found, the *check rotation symmetric* subroutine, CRTS, determines whether the rotation is symmetric. If so, $nsrtw$ is increased accordingly. Finally, the function returns the number of rotational symmetries found.

The Least Common Multiple subroutine, LCM, is used to find the number of rotations nrt represented by an AAC map $aacm$ and number of points np.

DRTS($np, pt[\,]$**)**
  Input:
    • number of points in the AAC, np
    • AAC point array, $pt[\,]$
  Output:
    • number of symmetric rotations found, $drts$
  Working:
    • AAC centroid, $cntrd$
    • number of symmetric rotations, working, $nsrtw$
    • AAC map, $aacm$
    • number of rotations, nrt

```
   // calculate centroid of AAC
cntrd = CNTRD(np, pt[ ])
   // initialize the working number of symmetric rotations
nsrtw = 1
   // for all AAC maps from 1 to (np − 1) ...
for aacm = 1 to (np − 1)
      // calculate the number of rotations
   nrt = LCM(np, aacm) / aacm
      // if the number of rotations cannot increase
      //    the working number of symmetric rotations,
      //    don't check the rotation for symmetry
   if nrt <= nsrtw then
    | continue
      // if the rotation is symmetric, update the working
      //    number of symmetric rotations
   if CRTS(np, pt[ ], cntrd, aacm, nrt) = TRUE then
    | nsrtw = nrt
   // return the number of symmetric rotations found
drts = nsrtw
```

Figure 6.2   Determine rotational symmetry function

6.2.2 CNTRD function

The CNTRD function calculates the centroid of a closed AAC. The centroid is a point in the xy plane. The x-coordinate of the centroid is the arithmetic mean of the x-coordinates of all points in the AAC; the y-coordinate of the centroid is calculated similarly. The CNTRD function is provided in Figure 6.3.

CNTRD($np, pt[\ ]$)
  Input:
     • number of points in the AAC, np
     • AAC point array, $pt[\ ]$
  Output:
     • centroid of the AAC, $cntrd$

$$cntrd.x = \frac{\sum\limits_{i=0}^{np-1} pt[i].x}{np}$$
$$cntrd.y = \frac{\sum\limits_{i=0}^{np-1} pt[i].y}{np}$$

Figure 6.3 Centroid function

6.2.3 CRTS function

The *Check Rotation Symmetric* function, CRTS, determines whether a rotation is symmetric for an AAC. Pseudocode for the CRTS function is provided in Figure 6.4 on the next page. The AAC point map and rotation angle to be tested for symmetry are specified by the input parameters *aacm* and *nrt*, respectively. The function returns $TRUE$ if the rotation is symmetric, and $FALSE$ otherwise.

The function rotates each point in the AAC in sequence until it either encounters a rotated point which is unequal to its corresponding mapped point, in which case the function declares the rotation asymmetric, or all rotated points in the AAC are found to be equal to their corresponding mapped points, in which case the function declares the rotation symmetric.

CRTS($np, pt[\,], cntrd, aacm, nrt$**)**

  Input:
- number of points in the AAC, np
- AAC point array, $pt[\,]$
- AAC centroid, $cntrd$
- AAC map, $aacm$
- number of rotations, nrt

  Output:
- TRUE if rotation is symmetric; FALSE otherwise

  Working:
- original point number, q
- rotated point, r
- mapped point number, s
- cosine, sine of rotation angle, cs, sn

```
// set the mapped point number for point 0
```
$s = aacm$
```
// calculate cosine & sine of rotation angle
```
$cs = \cos(\frac{2\pi}{nrt}); \; sn = \sin(\frac{2\pi}{nrt})$
```
// for all points in the AAC ...
```
for $q = 0$ **to** $(np - 1)$
```
    // calculate rotated point
```
    $r.x = cs * (pt[q].x - cntrd.x) - sn * (pt[q].y - cntrd.y) + cntrd.x$
    $r.y = sn * (pt[q].x - cntrd.x) + cs * (pt[q].y - cntrd.y) + cntrd.y$
```
    // if rotated point is unequal to mapped point, the
    //     rotation is not symmetric
```
    if $(r.x \neq pt[s].x)$ **or** $(r.y \neq pt[s].y)$ **then**
      | **return** *FALSE*
```
    // find the mapped point number for point q + 1
```
    $s = (s + 1) \bmod np$
```
// rotation is symmetric
```
return *TRUE*

Figure 6.4   Check rotation symmetric function

6.3 DRTS example

This section presents an example of DRTS algorithm operation. The section begins with a graph of the example AAC, followed by a trace showing how the algorithm determines the number of rotational symmetries of the AAC. The section ends with a discussion of the trace.

6.3.1 Graph

Figure 6.5 on the following page shows an order 3 AAC with five rotational symmetries.

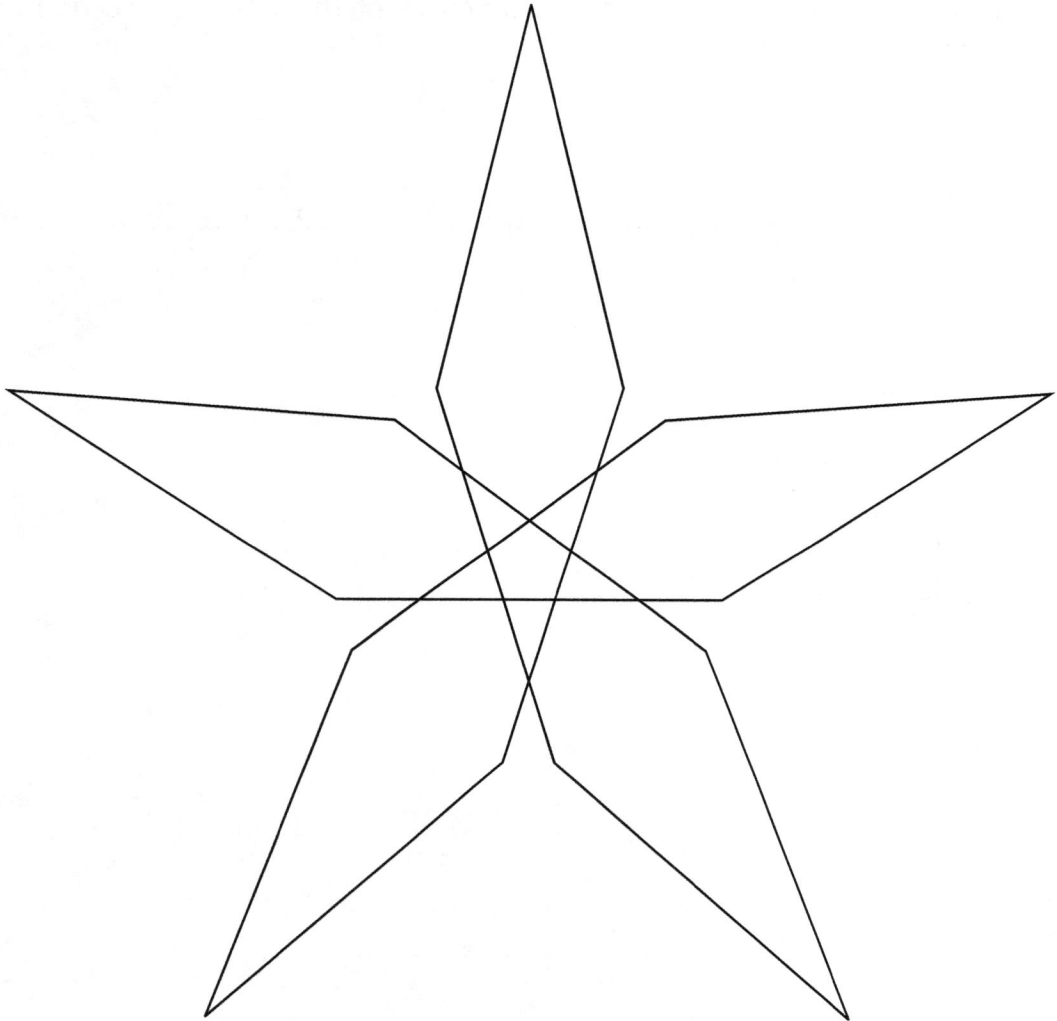

Figure 6.5   AAC [45: 5 4 0 15]

6.3.2 Trace

The DRTS algorithm determines the number of rotational symmetries of
AAC [45: 5 4 0 15] as follows:

```
DTRS:
  np = 15
  pt[0] - pt[14]:
    pt[ 0] = (10.000 10.000)
    pt[ 1] = (17.660 16.428)
    pt[ 2] = (20.751 25.938)
    pt[ 3] = (18.331 35.641)
    pt[ 4] = (15.912 25.938)
    pt[ 5] = (19.002 16.428)
    pt[ 6] = (26.663 10.000)
    pt[ 7] = (22.917 19.272)
    pt[ 8] = (14.827 25.150)
    pt[ 9] = ( 4.851 25.847)
    pt[10] = (13.331 20.548)
    pt[11] = (23.331 20.548)
    pt[12] = (31.812 25.847)
    pt[13] = (21.836 25.150)
    pt[14] = (13.746 19.272)

  cntrd = (18.331 21.467)
  nsrtw = 1
  for aacm = 1 to 14:
    aacm = 1:
      nrt = LCM(15,1) / 1 = 15
      CRTS:
        for q = 0 to 14:
          q =  0 : r = (15.384  7.603) <> pt[1]
    aacm = 2:
      nrt = LCM(15,2) / 2 = 15
      CRTS:
        for q = 0 to 14:
          q =  0 : r = (15.384  7.603) <> pt[2]
```

```
aacm = 3:
  nrt = LCM(15,3) / 3 = 5
  CRTS:
    for q = 0 to 14:
      q =  0 : r = (26.663 10.000) <> pt[3]
aacm = 4:
  nrt = LCM(15,4) / 4 = 15
  CRTS:
    for q = 0 to 14:
      q =  0 : r = (15.384  7.603) <> pt[4]
aacm = 5:
  nrt = LCM(15,5) / 5 = 3
  CRTS:
    for q = 0 to 14:
      q =  0 : r = (32.428 19.986) <> pt[5]
aacm = 6:
  nrt = LCM(15,6) / 6 = 5
  CRTS:
    for q = 0 to 14:
      q =  0 : r = (26.663 10.000) = pt[6]
      q =  1 : r = (22.917 19.272) = pt[7]
      q =  2 : r = (14.827 25.150) = pt[8]
      q =  3 : r = ( 4.851 25.847) = pt[9]
      q =  4 : r = (13.331 20.548) = pt[10]
      q =  5 : r = (23.331 20.548) = pt[11]
      q =  6 : r = (31.812 25.847) = pt[12]
      q =  7 : r = (21.836 25.150) = pt[13]
      q =  8 : r = (13.746 19.272) = pt[14]
      q =  9 : r = (10.000 10.000) = pt[0]
      q = 10 : r = (17.660 16.428) = pt[1]
      q = 11 : r = (20.751 25.938) = pt[2]
      q = 12 : r = (18.331 35.641) = pt[3]
      q = 13 : r = (15.912 25.938) = pt[4]
      q = 14 : r = (19.002 16.428) = pt[5]
  nsrtw = 5
```

```
aacm = 7:
  nrt = LCM(15,7) / 7 = 15
  CRTS:
    for q = 0 to 14:
      q =  0 : r = (15.384   7.603) <> pt[7]
aacm = 8:
  nrt = LCM(15,8) / 8 = 15
  CRTS:
    for q = 0 to 14:
      q =  0 : r = (15.384   7.603) <> pt[8]
aacm = 9:
  nrt = LCM(15,9) / 9 = 5
aacm = 10:
  nrt = LCM(15,10) / 10 = 3
aacm = 11:
  nrt = LCM(15,11) / 11 = 15
  CRTS:
    for q = 0 to 14:
      q =  0 : r = (15.384   7.603) <> pt[11]
aacm = 12:
  nrt = LCM(15,12) / 12 = 5
aacm = 13:
  nrt = LCM(15,13) / 13 = 15
  CRTS:
    for q = 0 to 14:
      q =  0 : r = (15.384   7.603) <> pt[13]
aacm = 14:
  nrt = LCM(15,14) / 14 = 15
  CRTS:
    for q = 0 to 14:
      q =  0 : r = (15.384   7.603) <> pt[14]

drts = 5
```

6.3.3 Discussion

As described in section 6.1, as it runs, the DRTS function performs symmetry tests only on those AAC maps that may increase the number of rotational symmetries found so far. The trace for this example illustrates this selective symmetry testing. The function processes the 14 AAC maps as follows:

· DRTS determines that AAC maps 1 through 5 are not symmetric rotations.

· DRTS determines that AAC map 6 is a symmetric rotation, and increases the working number of rotational symmetries to 5.

· DRTS determines that AAC maps 7 and 8 are not symmetric rotations.

· DRTS skips testing AAC maps 9 and 10 for symmetry, because neither AAC map has the potential to increase the number of rotational symmetries found so far. AAC map 9 represents 5 rotations and AAC map 10 represents 3 rotations.

· DRTS determines that AAC map 11 is a not symmetric rotation.

· DRTS skips testing AAC map 12 for symmetry, because this AAC map represents 5 rotations and thus does not have the potential to increase the number of rotational symmetries found so far.

· DRTS determines that AAC maps 13 and 14 are not symmetric rotations.

The function concludes that the AAC has 5 rotational symmetries.

7 Determining Accelerating Angle Curve Reflectional Symmetry

This chapter provides a method of determining the number of reflectional symmetries (DRFS) of a closed AAC given only the AAC's point data. The first four sections provide technical background for the DRFS algorithm. These sections describe a curve partition, describe how the DRFS algorithm generates a partition axis and tests whether it is a reflection axis, and provide examples of the different types of reflection axis. The fifth section describes the DRFS algorithm. The last two sections provide examples of DRFS algorithm operation. The first example demonstrates DRFS algorithm operation on an AAC with an even number of points. In the second example, the AAC has an odd number of points.

7.1 Curve partition

A curve partition divides an AAC into two open subcurves. There are four types of curve partition:

1. *<bisect-line-segment bisect-line-segment>* (<BLS BLS>)

2. *<intersect-point intersect-point>* (<IPT IPT>)

3. *<bisect-line-segment intersect-point>* (<BLS IPT>)

4. *<intersect-point bisect-line_segment>* (<IPT BLS>)

Figure 7.1 on page 135 provides an example of a <BLS BLS> curve partition. The two subcurves consist of points $< P\ 0\ 7\ 6\ 5\ Q >$ and $< P\ 1\ 2\ 3\ 4\ Q >$. Point P is the midpoint of the line segment joining points 0 and 1. Point Q is the midpoint of the line segment joining points 5 and 4. The curve partition is represented by an ordered list of ordered pairs: $<< 0\ 1 >< 7\ 2 >< 6\ 3 >< 5\ 4 >>$.

133

Figure 7.2 on page 136 provides an example of an <IPT IPT> curve partition. The two subcurves consist of points $< 0\ 7\ 6\ 5\ 4 >$ and $< 0\ 1\ 2\ 3\ 4 >$. The curve partition is represented as $<< 0\ 0 >< 7\ 1 >< 6\ 2 >< 5\ 3 >< 4\ 4 >>$.

Figure 7.3 on page 137 provides an example of a <BLS IPT> curve partition. The two subcurves consist of points $< P\ 0\ 8\ 7\ 6\ 5 >$ and $< P\ 1\ 2\ 3\ 4\ 5 >$. The curve partition is represented as $<< 0\ 1 >< 8\ 2 >< 7\ 3 >< 6\ 4 >< 5\ 5 >>$.

Figure 7.4 on page 138 provides an example of an <IPT BLS> curve partition. The two subcurves consist of points $< 0\ 8\ 7\ 6\ 5\ Q >$ and $< 0\ 1\ 2\ 3\ 4\ Q >$. The curve partition is represented as $<< 0\ 0 >< 8\ 1 >< 7\ 2 >< 6\ 3 >< 5\ 4 >>$.

A curve partition can be specified implicitly by providing an initial ordered pair $< a\ b >$ and the number of ordered pairs, nop. The remaining ordered pairs can be generated as:

$$< ((a - 1) \bmod np)\ ((b + 1) \bmod np) >$$
$$< ((a - 2) \bmod np)\ ((b + 2) \bmod np) >$$
$$\cdots$$
$$< ((a - (nop - 1)) \bmod np)\ ((b + (nop - 1)) \bmod np) > .$$

The total number of curve partitions in a closed AAC is equal to its number of points, np.

When the number of points in an AAC is even, the AAC has $\frac{np}{2}$ <BLS BLS> curve partitions and $\frac{np}{2}$ <IPT IPT> curve partitions. There are $\frac{np}{2}$ ordered pairs in a <BLS BLS> curve partition and $\frac{np}{2} + 1$ ordered pairs in an <IPT IPT> curve partition.

When the number of points in an AAC is odd, the DRFS algorithm examines $\lfloor \frac{np}{2} \rfloor$ <BLS IPT> curve partitions and $\lfloor \frac{np}{2} \rfloor + 1$ <IPT BLS> curve partitions.[1] There are $\lfloor \frac{np}{2} \rfloor + 1$ ordered pairs in a <BLS IPT> curve partition and $\lfloor \frac{np}{2} \rfloor + 1$ ordered pairs in an <IPT BLS> curve partition.

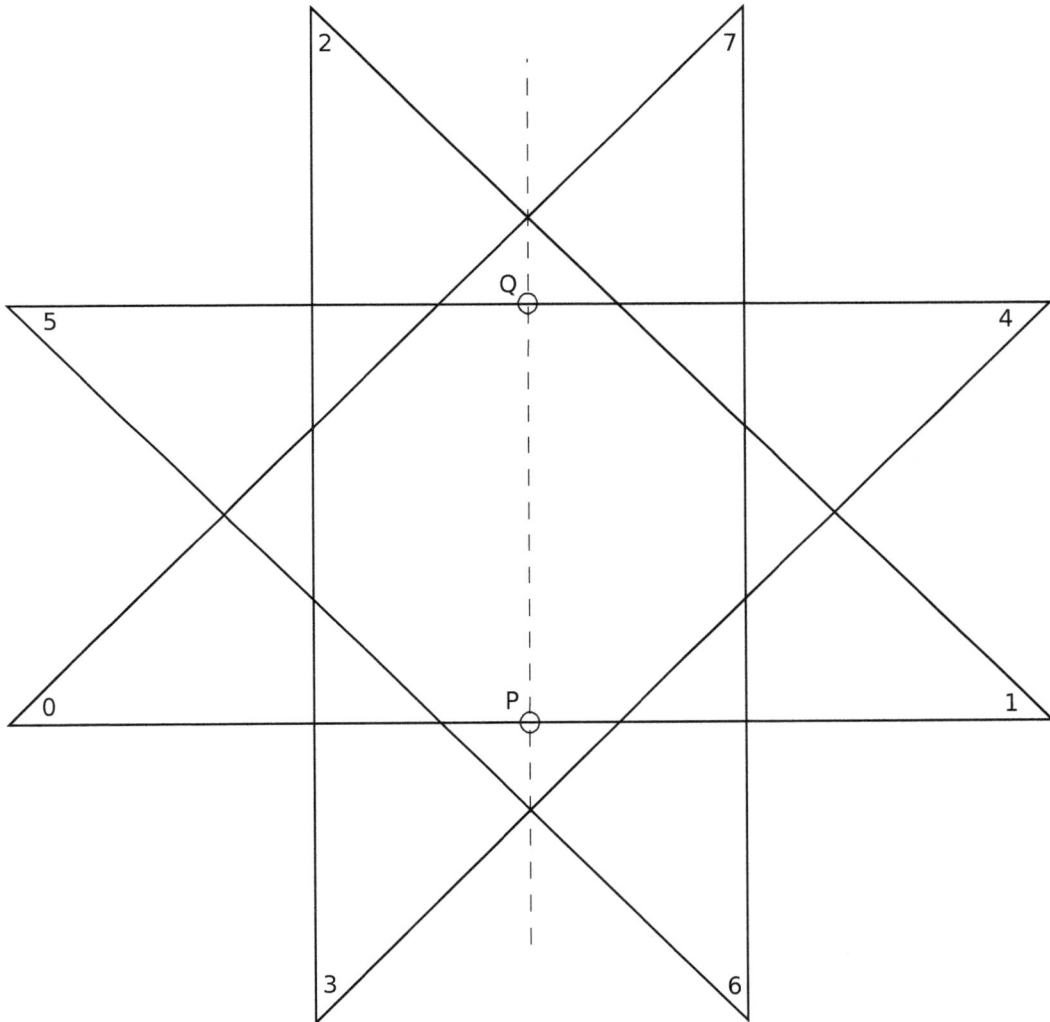

Figure 7.1 <BLS BLS> curve partition

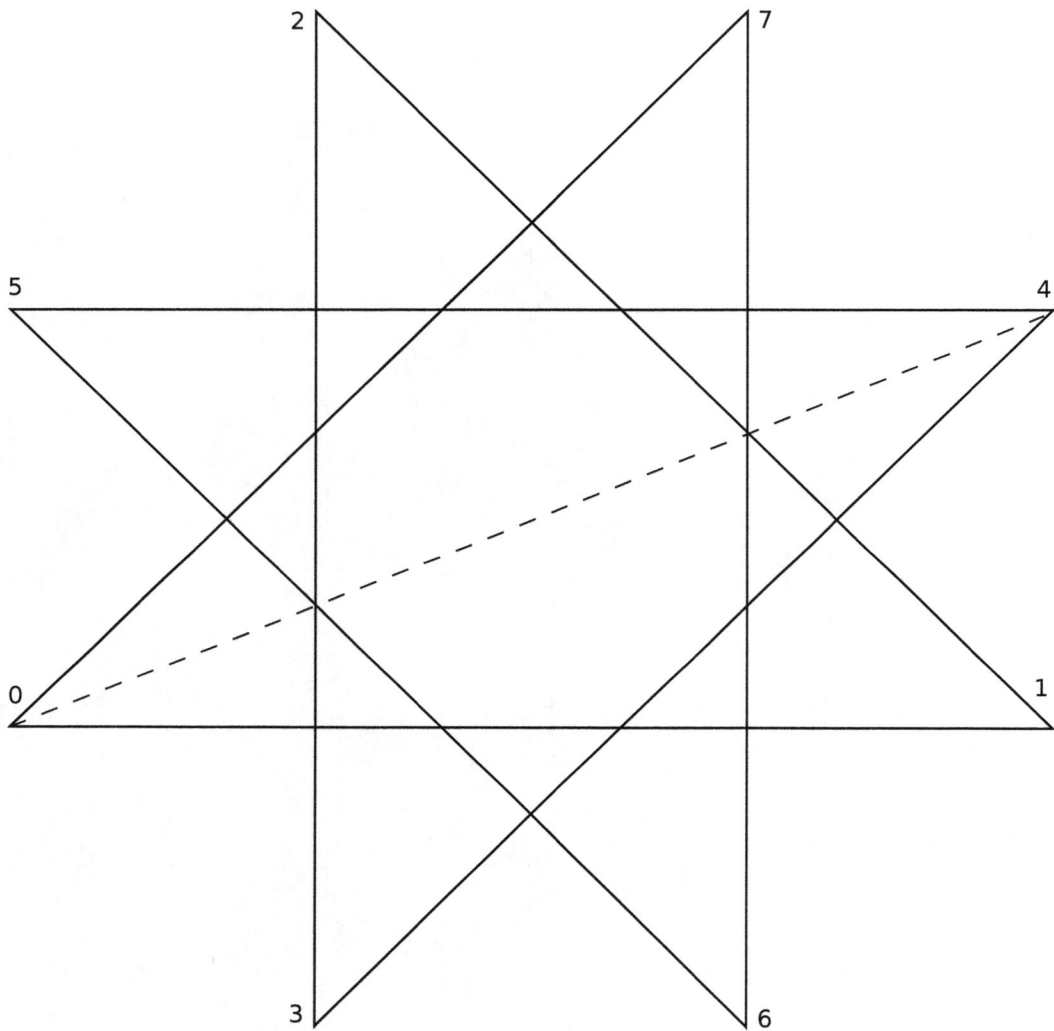

Figure 7.2   <IPT IPT> curve partition

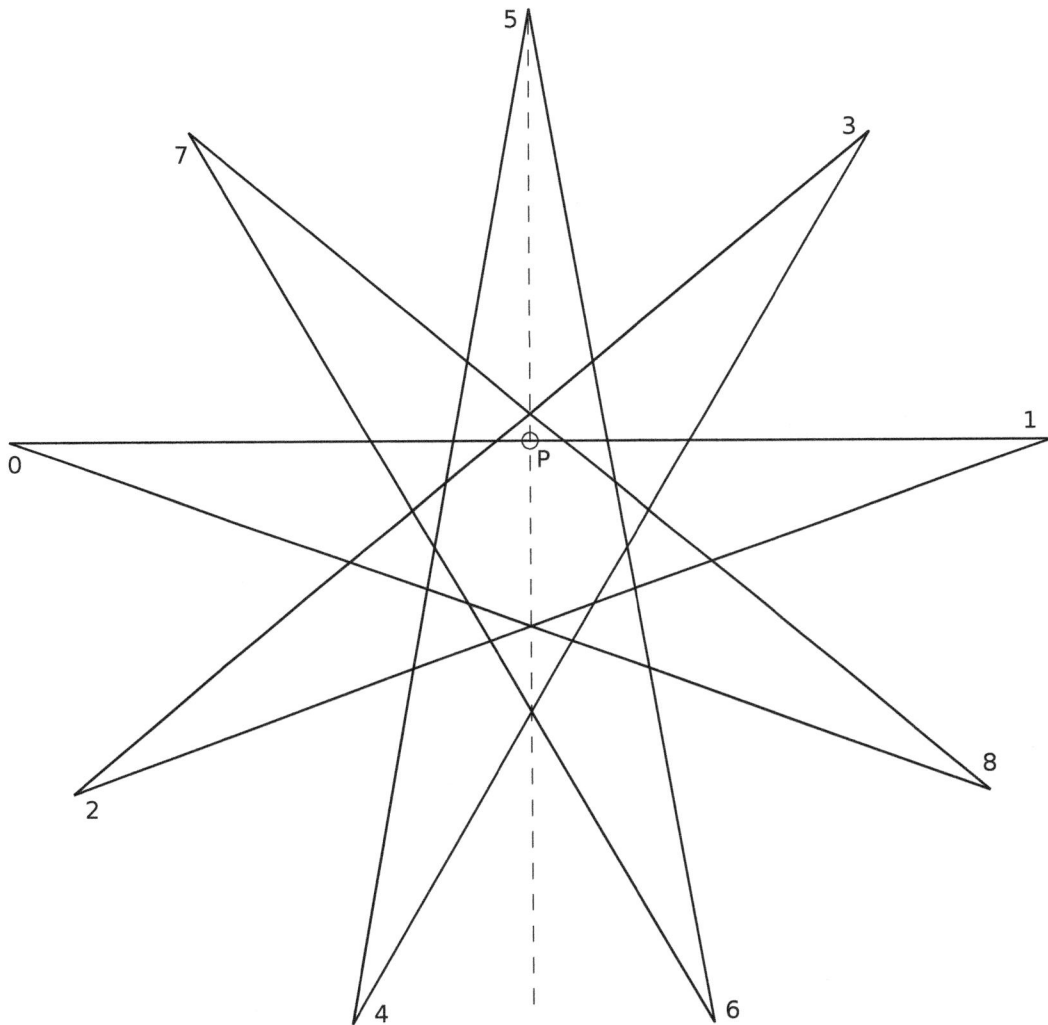

Figure 7.3   <BLS IPT> curve partition

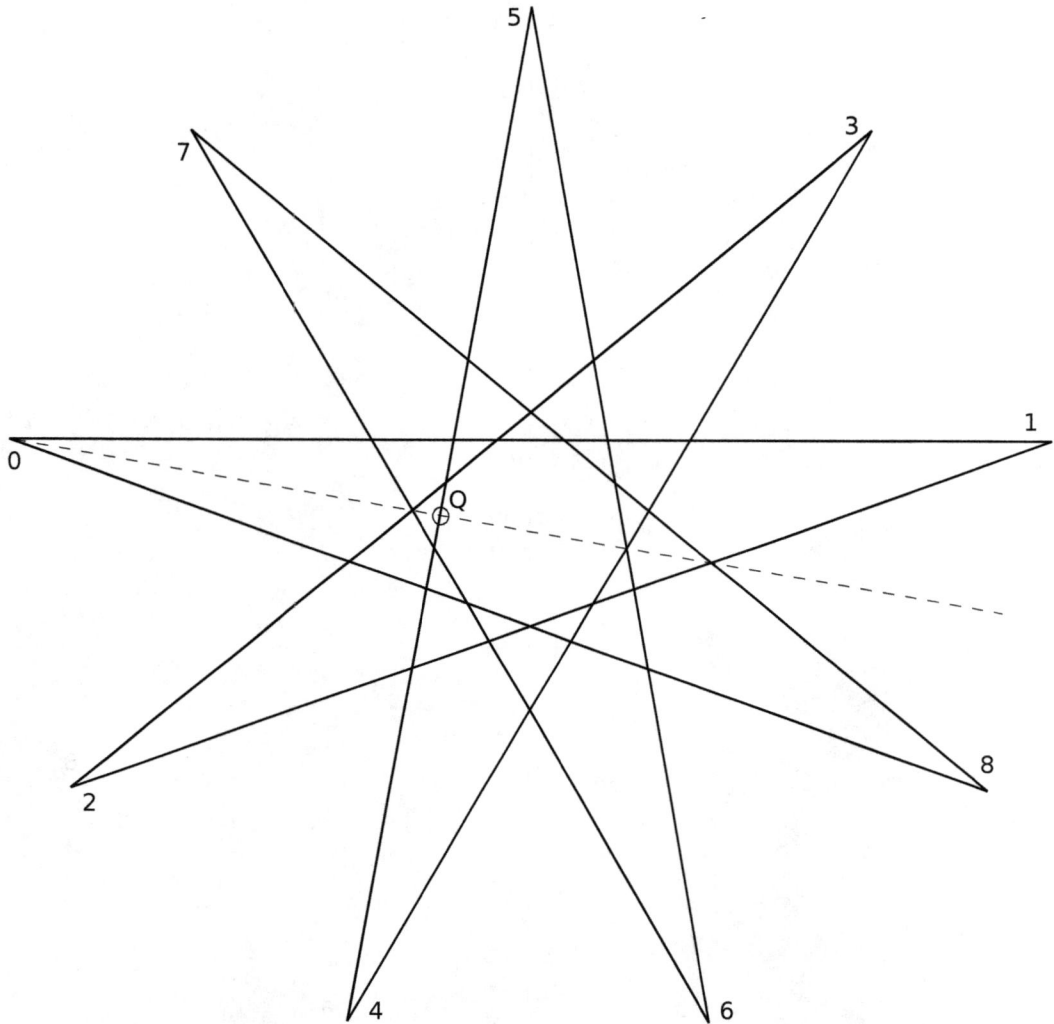

Figure 7.4 <IPT BLS> curve partition

7.2   Finding a partition axis

In order to test whether a curve partition is reflectional, the DRFS algorithm generates a line intersecting the AAC called a *partition axis*. A partition axis is defined by two points, c and d. When a curve partition is non-reflectional, the partition axis depends upon the particular ordered pairs used to generate it. In contrast, when a curve partition is reflectional, the partition axis is unique and is called a *reflection axis*. Reflection axes are discussed further in section 7.4 on page 144.

The DRFS algorithm employs two methods of generating a partition axis: the *midpoint method* and the *circle method*. The midpoint method is tried first because it is most efficient; if this method fails, then the circle method is employed. The next two subsections describe the midpoint and circle methods.

7.2.1   Midpoint method

The midpoint method of calculating a partition axis consists of two steps:
(1) Point c is calculated as the midpoint of the line segment joining the points in the first ordered pair within the curve partition.
(2) The remaining ordered pairs in the curve partition are searched for an ordered pair with midpoint m unequal to point c. When such an ordered pair is found, point d is set to point m. If no such ordered pair is found, the midpoint method fails.

Figure 7.5 on the following page provides three examples of midpoint method operation:

(1) The <BLS BLS> curve partition is $<< 0\ 1 >< 5\ 2 >< 4\ 3 >>$. Point c is the midpoint of the line segment joining points 0 and 1, and point d is midpoint of the line segment joining points 5 and 2, resulting in the partition axis shown.

(2) The <IPT IPT> curve partition is $<< 0\ 0 >< 5\ 1 >< 4\ 2 >< 3\ 3 >>$. Point c is the midpoint of the line segment joining points 0 and 0 (i.e., just point 0), and point d is midpoint of the line segment joining points 5 and 1, resulting in the partition axis shown.

(3) The curve in this example doubles back upon itself. The <IPT IPT> curve partition is << 0 0 >< 7 1 >< 6 2 >< 5 3 >< 4 4 >>. Point c is the midpoint of the line segment joining points 0 and 0 (i.e., just point 0). The midpoints of the line segments joining the points in all other ordered pairs in the curve partition are equal to point c, so the midpoint method fails to identify point d and thus no partition axis is found.

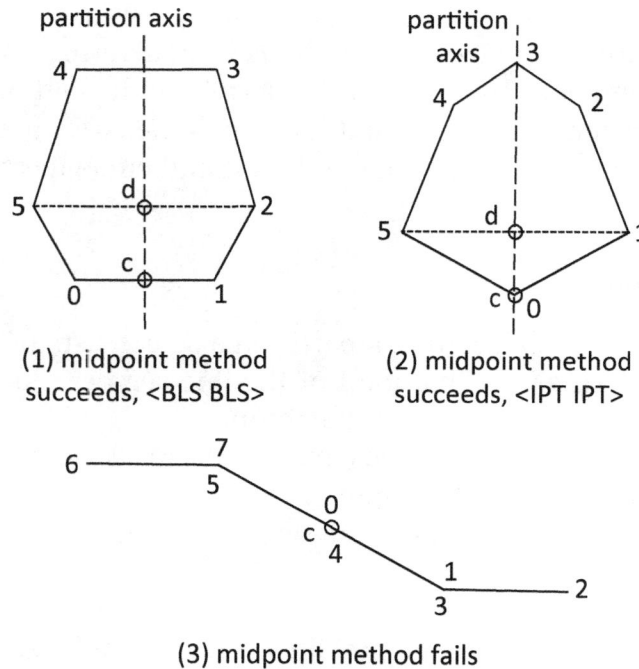

Figure 7.5 Midpoint method examples

7.2.2 Circle method

The circle method of calculating a partition axis consists of four steps:
(1) Search the ordered pairs in the curve partition for an ordered pair containing noncoincident points. The two points define a line segment ls of nonzero length.
(2) Calculate the midpoint m of line segment ls.
(3) Center a circle of radius r at point m.

(4) Calculate partition axis points c and d as the two points where a line drawn perpendicular to line segment ls through point m intersects the circle.

Figure 7.6 provides two examples of circle method operation:

(1) The curve partition is $<< 0\ 0 >< 5\ 1 >< 4\ 2 >< 3\ 3 >>$. The first ordered pair in the curve partition containing noncoincident points is $< 5\ 1 >$. A circle of radius r is centered on the midpoint m of the line segment ls joining points 5 and 1. Points c and d are calculated as the points where the line perpendicular to ls through midpoint m intersects the circle.

(2) The curve partition is $<< 0\ 0 >< 7\ 1 >< 6\ 2 >< 5\ 3 >< 4\ 4 >>$. The first ordered pair in the curve partition containing noncoincident points is $< 7\ 1 >$. A circle of radius r is centered on the midpoint m of the line segment ls joining points 7 and 1. Points c and d are calculated as the points where the line perpendicular to ls through midpoint m intersects the circle.

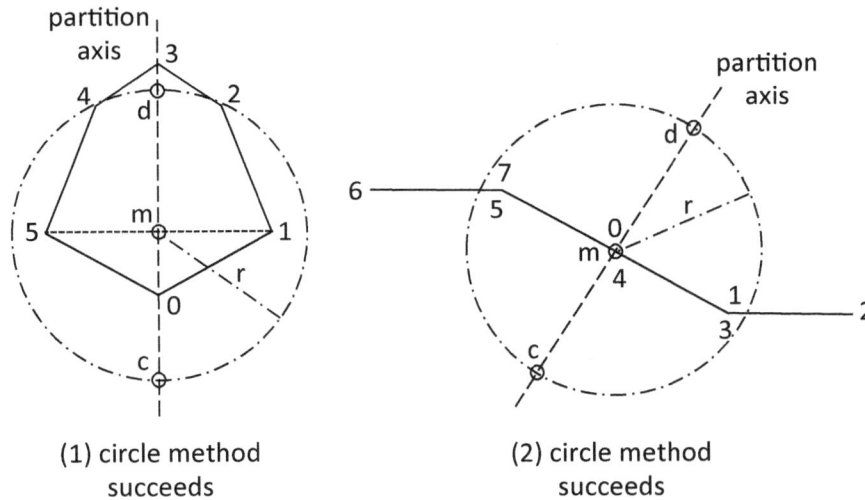

(1) circle method succeeds

(2) circle method succeeds

Figure 7.6 Circle method examples

7.3 Testing whether a partition axis is a reflection axis

A partition axis is a reflection axis if all of the associated curve partition's ordered pairs pass both the *midpoint test* and the *projection test*. The midpoint test determines whether the midpoint of the line segment joining the two points in an ordered pair lies on the partition axis. The projection test determines whether the two points in an ordered pair project onto the same point on the partition axis. Each point is projected onto the partition axis via a line perpendicular to the axis.

Figure 7.7 on the facing page shows the four possible outcomes from partition axis reflection testing for ordered pair $< a\ b >$. These outcomes are:

(1) the midpoint m of the line segment joining points a and b lies on the partition axis (the midpoint test passes), and points a and b project onto the same point on the partition axis (the projection test passes). The projections of points a and b are labeled as points a' and b', respectively, in the figure.

(2) the midpoint m of the line segment joining points a and b lies off the partition axis (the midpoint test fails), and points a and b project onto the same point on the partition axis (the projection test passes).

(3) the midpoint m of the line segment joining points a and b lies on the partition axis (the midpoint test passes), and points a and b project onto different points on the partition axis (the projection test fails).

(4) the midpoint m of the line segment joining points a and b lies off the partition axis (the midpoint test fails), and points a and b project onto different points on the partition axis (the projection test fails).

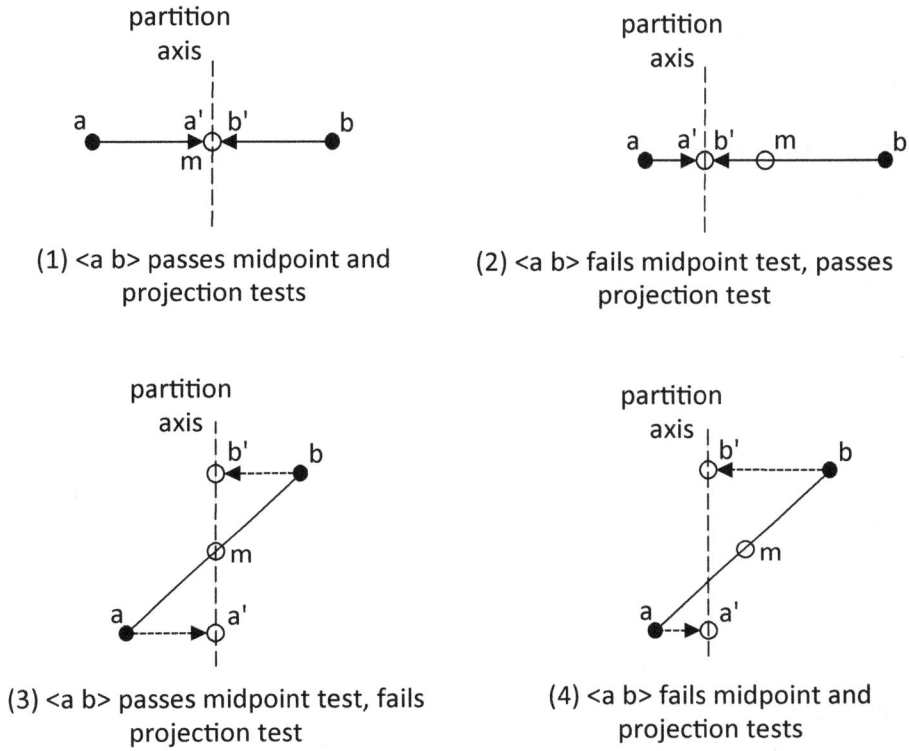

(1) <a b> passes midpoint and projection tests

(2) <a b> fails midpoint test, passes projection test

(3) <a b> passes midpoint test, fails projection test

(4) <a b> fails midpoint and projection tests

Figure 7.7 Reflection test outcomes on an ordered pair

7.4   Reflection axis types

The following four figures provide examples of the four types of reflection axis and the AACs in which they appear:

- Figure 7.8 on the next page shows AAC [12: 1 1 4 7], which has three <BLS BLS> and zero <IPT IPT> reflection axes.

- Figure 7.9 on page 146 shows AAC [8: 1 5 1 2], which has four <IPT IPT> and zero <BLS BLS> reflection axes.

- Figure 7.10 on page 147 shows AAC [60: 4 2 28 56], which has three <BLS BLS> and three <IPT IPT> reflection axes.

- Figure 7.11 on page 148 shows AAC [45: 7 4 30 15], which has five reflection axes. The DRFS algorithm divides these axes into two <BLS IPT> reflection axes and three <IPT BLS> reflection axes.

As defined in Chapter 4, when an AAC is rotated 180 degrees through the z dimension around a reflection axis, the rotated AAC traces the same path in the xy plane as the original AAC, but in the opposite direction. This path direction reversal can be verified using the ordered pairs in a reflectional curve partition. For example, for the AAC in Figure 7.1 on page 135, the original path traverses the points in numerical order: $< 0 1 2 3 4 5 6 7 0 >$. The curve partition shown is $<< 0 1 >< 7 2 >< 6 3 >< 5 4 >>$. The reflected path can be generated from the original path by treating the curve partition's ordered pairs as swap instructions: swap point 0 and point 1, point 7 and point 2, point 6 and point 3, and point 5 and point 4. The resulting reflected path traverses the AAC's points in reverse order: $< 1 0 7 6 5 4 3 2 1 >$.

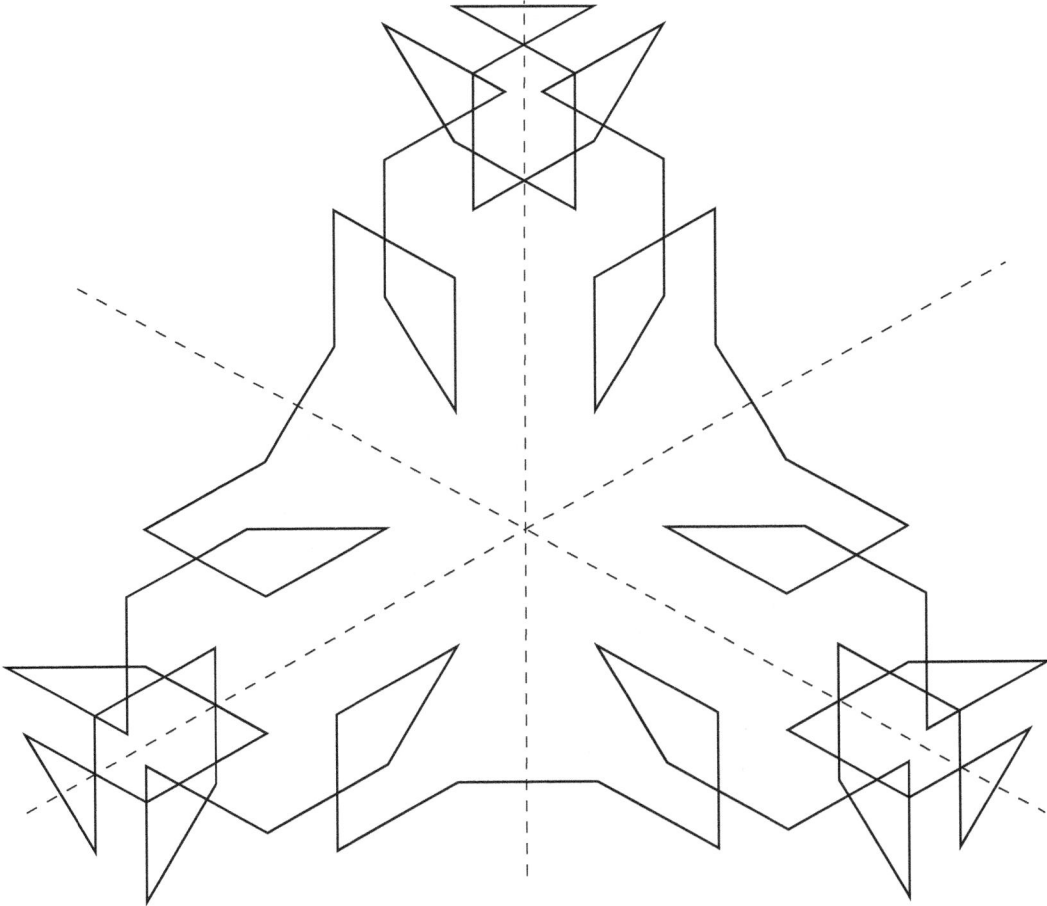

Figure 7.8 AAC with only <BLS BLS> reflection axes

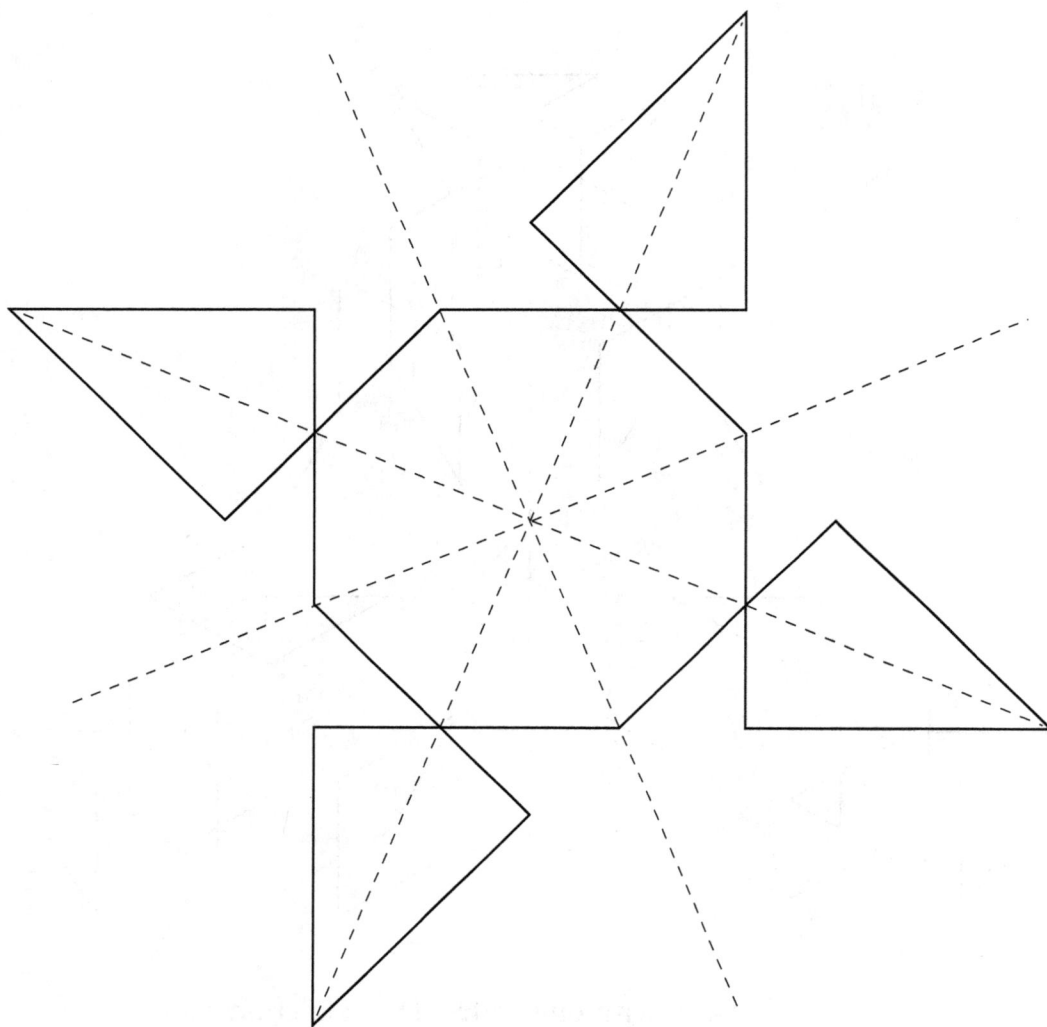

Figure 7.9 AAC with only <IPT IPT> reflection axes

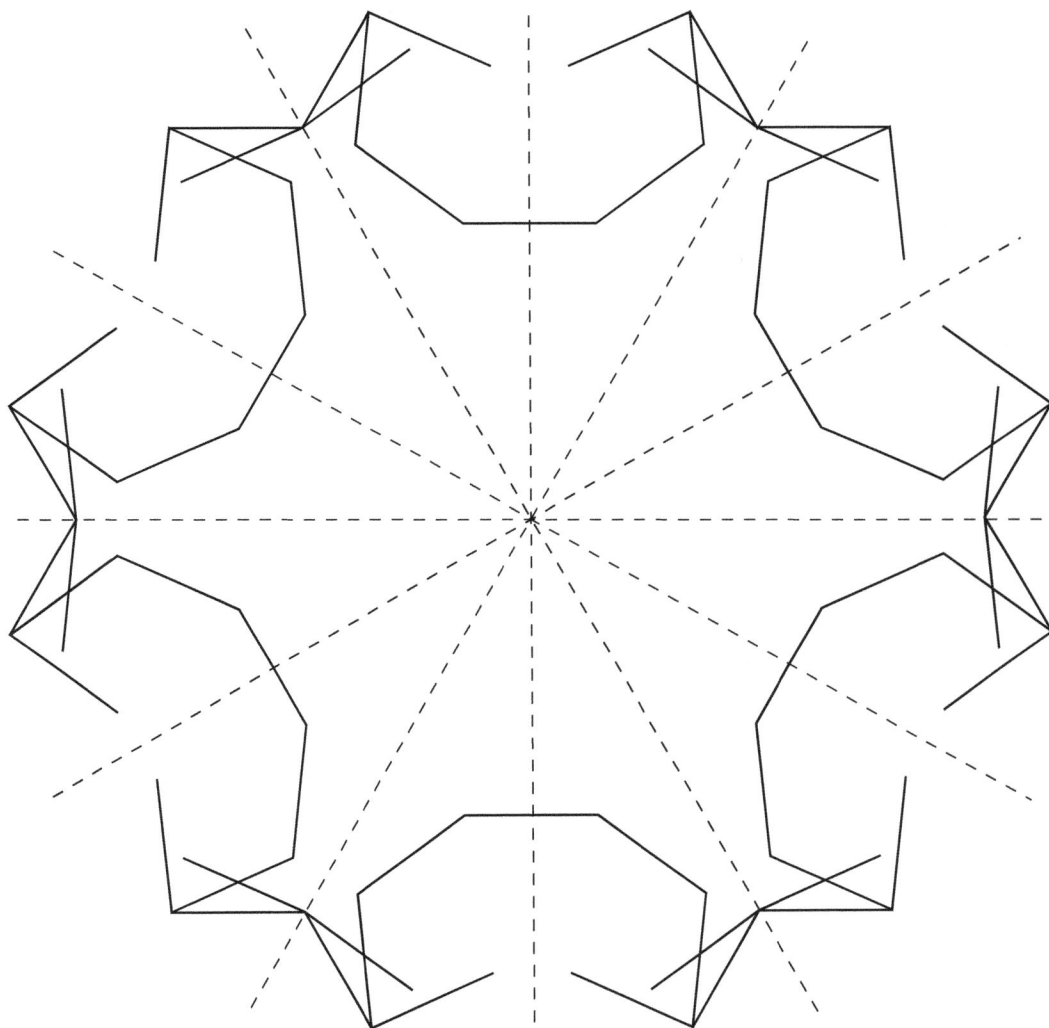

Figure 7.10 AAC with both <BLS BLS> and <IPT IPT> reflection axes

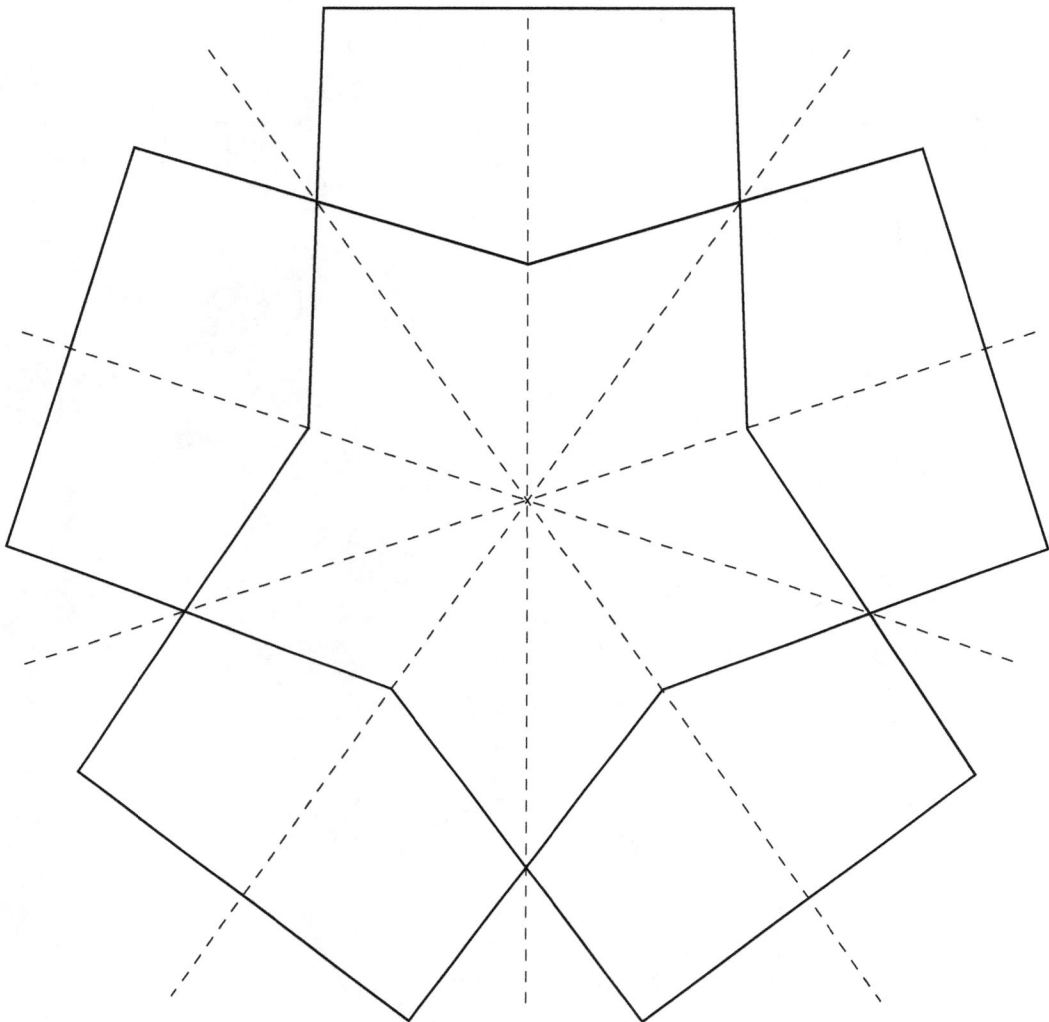

Figure 7.11 AAC with both <BLS IPT> and <IPT BLS> reflection axes

7.5   DRFS algorithm

This section presents an algorithm for determining the number of reflectional symmetries of an AAC given only its point data. The first subsection describes the DRFS algorithm main routine, DRFS. The second subsection describes the DRFS subroutine CCPRF, which checks whether a curve partition is reflectional.

7.5.1   DRFS main routine

Function DRFS is shown in Figure 7.12 on the following page. The DRFS function can be roughly divided into two parts, one which executes when the number of AAC points is even, and the other which executes when the number of AAC points is odd. When the number of AAC points is even, <BLS BLS> and <IPT IPT> curve partitions are searched for reflectional curve partitions. When the number of AAC points is odd, <BLS IPT> and <IPT BLS> curve partitions are searched. Each curve partition is tested by a call to subroutine CCPRF. The DRFS function concludes by reporting the number of reflectional symmetries of the AAC as the number of reflectional curve partitions found during its search.

The working variables employed by the DRFS function are:
- number of reflectional curve partitions found, working, $nrfcpw$
- number of curve partitions to search, ncp
- number of ordered pairs in the curve partition, nop
- curve partition number, cpn

DRFS($np, pt[\ ]$)
 Input:
  • number of points in the AAC, np
  • AAC point array, $pt[\ ]$
 Output:
  • number of reflectional symmetries found, $drfs$

```
// initialize working number of reflectional curve partitions found
```
$nrfcpw = 0$
if $np \bmod 2 = 0$ **then**   `// if number of AAC points is even ...`
  `// --- search <BLS BLS> curve partitions ---`
  $ncp = np/2;\ nop = np/2$
  for $cpn = 0$ **to** $(ncp - 1)$
   if CCPRF$(np, pt, < cpn\ (cpn + 1) >, nop) = TRUE$ **then**
    $nrfcpw = nrfcpw + 1$
  `// --- search <IPT IPT> curve partitions ---`
  $ncp = np/2;\ nop = np/2 + 1$
  for $cpn = 0$ **to** $(ncp - 1)$
   if CCPRF$(np, pt, < cpn\ cpn >, nop) = TRUE$ **then**
    $nrfcpw = nrfcpw + 1$
else   `// number of AAC points is odd ...`
  `// --- search <BLS IPT> curve partitions ---`
  $ncp = np/2;\ nop = np/2 + 1$
  for $cpn = 0$ **to** $(ncp - 1)$
   if CCPRF$(np, pt, < cpn\ (cpn + 1) >, nop) = TRUE$ **then**
    $nrfcpw = nrfcpw + 1$
  `// --- search <IPT BLS> curve partitions ---`
  $ncp = np/2 + 1;\ nop = np/2 + 1$
  for $cpn = 0$ **to** $(ncp - 1)$
   if CCPRF$(np, pt, < cpn\ cpn >, nop) = TRUE$ **then**
    $nrfcpw = nrfcpw + 1$
end
```
// return the number of reflectional curve partitions found
```
$drfs = nrfcpw$

Figure 7.12 Determine reflectional symmetry function

7.5.2 CCPRF subroutine

The DRFS subroutine *check curve partition reflectional*, CCPRF, is shown in Figure 7.13. CCPRF employs the implicit curve partition specification described in section 7.1, where the initial ordered pair in the curve partition, $< a\ b >$, and number of ordered pairs in the curve partition, nop, is provided.

CCPRF($np, pt[\ ], < a\ b >, nop$**)**
  Input:
    • number of points in the AAC, np
    • AAC point array, $pt[\ ]$
    • curve partition, initial ordered pair, $< a\ b >$
    • curve partition, number of ordered pairs, nop
  Output:
    • TRUE if curve partition is reflectional; FALSE otherwise

```
    // find points c and d defining a partition axis
```
    $ptc, ptd = \mathbf{FPA}(np, pt, < a\ b >, nop)$

```
    // if partition axis is reflectional ...
```
  if $\mathrm{CPARF}(np, pt, ptc, ptd, < a\ b >, nop) = TRUE$ **then**
```
        // report that the curve partition is reflectional
```
    return *TRUE*

```
    // report that the curve partition is non-reflectional
```
  return *FALSE*

Figure 7.13 Check curve partition reflectional function

CCPRF begins by calling subfunction *find partition axis*, FPA, to find points c and d defining a partition axis. The high-level operation of FPA is described in section 7.2 on page 139. CCPRF then checks whether the partition axis is reflectional by calling subfunction *check partition axis reflectional*, CPARF. The high-level operation of CPARF is described in section 7.3 on page 142. If the partition axis is found to be reflectional, then the curve partition is reported as reflectional. Otherwise, the curve partition is reported as non-reflectional.

Chapter 7

7.6   DRFS example 1

This section provides an example of the DRFS algorithm operating on closed AAC [20: 0 15 0 12], which has an even number of points. The example is divided into four parts: (1) a graph of the AAC; (2) a listing containing the AAC's point data; (3) a trace of DRFS algorithm operation; and (4) a discussion of the trace.

7.6.1   Graph

Figure 7.14 on the facing page shows that the AAC has four reflectional symmetries.

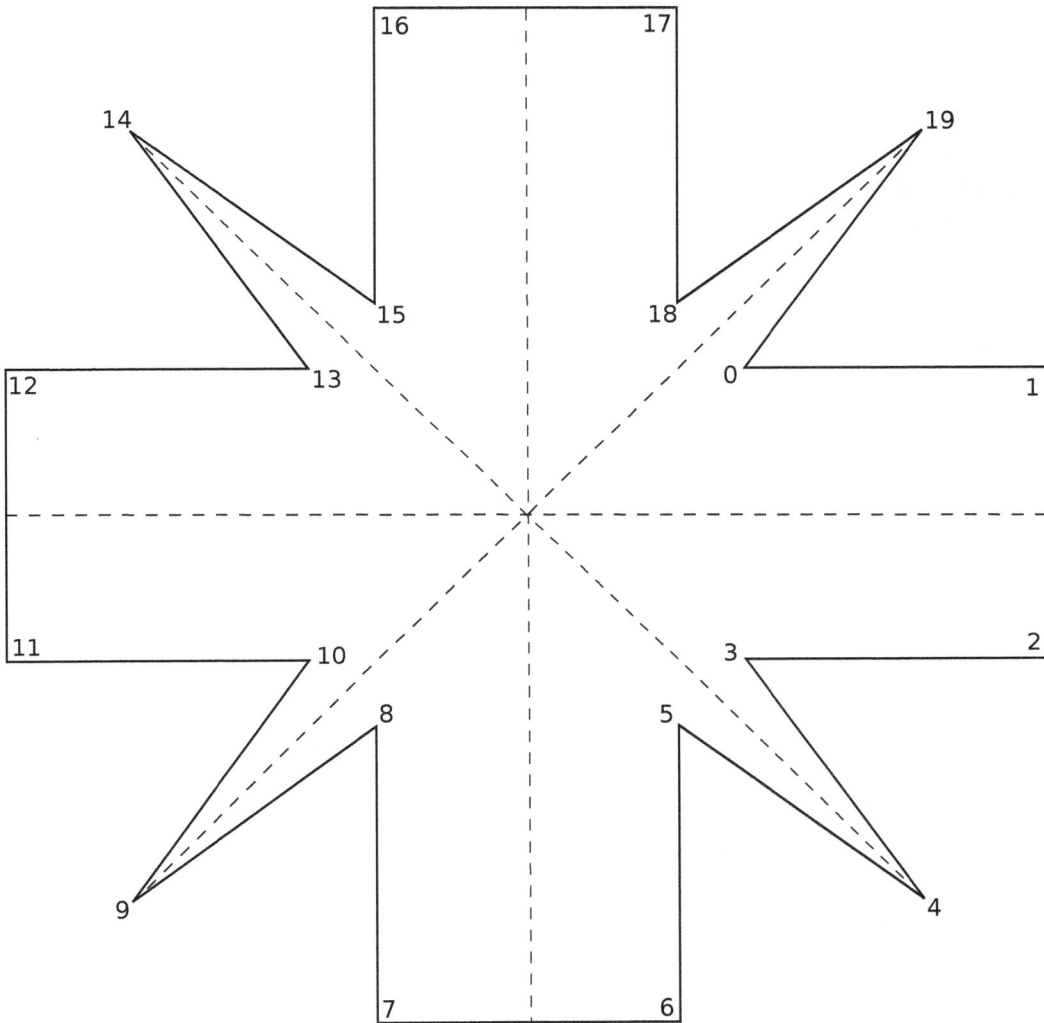

Figure 7.14   AAC [20: 0 15 0 12]

7.6.2   Point data

Listing 7.1 shows the point data for the AAC.

Listing 7.1

```
pt[ 0] = ( 10.000   10.000)
pt[ 1] = ( 20.000   10.000)
pt[ 2] = ( 20.000    0.000)
pt[ 3] = ( 10.000    0.000)
pt[ 4] = ( 15.878   -8.090)
pt[ 5] = (  7.788   -2.212)
pt[ 6] = (  7.788  -12.212)
pt[ 7] = ( -2.212  -12.212)
pt[ 8] = ( -2.212   -2.212)
pt[ 9] = (-10.302   -8.090)
pt[10] = ( -4.425    0.000)
pt[11] = (-14.425    0.000)
pt[12] = (-14.425   10.000)
pt[13] = ( -4.425   10.000)
pt[14] = (-10.302   18.090)
pt[15] = ( -2.212   12.212)
pt[16] = ( -2.212   22.212)
pt[17] = (  7.788   22.212)
pt[18] = (  7.788   12.212)
pt[19] = ( 15.878   18.090)
```

7.6.3 Trace

The DRFS algorithm determines the number of reflectional symmetries of the AAC as follows:

```
DRFS:
  >> testing 10 <BLS BLS> curve partitions <<

  == curve partition 0 test ==

  CCPRF:

    FPA:
      ordered pair 0: <0 1>
      point c = (15.000 10.000)

      ordered pair 1: <19 2>
      point d = (17.939 9.045)

    CPARF:
      == midpoint testing ==

      ordered pair 0: <0 1>
      point m = (15.000 10.000)
      dist m to axis = 0.000

      ordered pair 1: <19 2>
      point m = (17.939 9.045)
      dist m to axis = 0.000

      ordered pair 2: <18 3>
      point m = (8.894 6.106)
      dist m to axis = 5.590

      curve partition 0 is not reflectional
```

```
== curve partition 1 test ==

CCPRF:

  FPA:
    ordered pair 0: <1 2>
    point c = (20.000 5.000)

    ordered pair 1: <0 3>
    point d = (10.000 5.000)

  CPARF:
    == midpoint testing ==

    ordered pair 0: <1 2>
    point m = (20.000 5.000)
    dist m to axis = 0.000

    ordered pair 1: <0 3>
    point m = (10.000 5.000)
    dist m to axis = 0.000

    ordered pair 2: <19 4>
    point m = (15.878 5.000)
    dist m to axis = 0.000

    ordered pair 3: <18 5>
    point m = (7.788 5.000)
    dist m to axis = 0.000

    ordered pair 4: <17 6>
    point m = (7.788 5.000)
    dist m to axis = 0.000
```

```
ordered pair 5: <16 7>
point m = (-2.212 5.000)
dist m to axis = 0.000


ordered pair 6: <15 8>
point m = (-2.212 5.000)
dist m to axis = 0.000


ordered pair 7: <14 9>
point m = (-10.302 5.000)
dist m to axis = 0.000


ordered pair 8: <13 10>
point m = (-4.425 5.000)
dist m to axis = 0.000


ordered pair 9: <12 11>
point m = (-14.425 5.000)
dist m to axis = 0.000


== projection testing ==

ordered pair 0: <1 2>
point  1 proj = (20.000 5.000)
point  2 proj = (20.000 5.000)
proj dx dy    = (0.000 0.000)


ordered pair 1: <0 3>
point  0 proj = (10.000 5.000)
point  3 proj = (10.000 5.000
proj dx dy    = (0.000 0.000)


ordered pair 2: <19 4>
point 19 proj = (15.878 5.000)
point  4 proj = (15.878 5.000
proj dx dy    = (0.000 0.000)
```

```
ordered pair 3: <18 5>
point 18 proj = (7.788 5.000)
point  5 proj = (7.788 5.000
proj dx dy    = (0.000 0.000)


ordered pair 4: <17 6>
point 17 proj = (7.788 5.000)
point  6 proj = (7.788 5.000
proj dx dy    = (0.000 0.000)


ordered pair 5: <16 7>
point 16 proj = (-2.212 5.000)
point  7 proj = (-2.212 5.000
proj dx dy    = (0.000 0.000)


ordered pair 6: <15 8>
point 15 proj = (-2.212 5.000)
point  8 proj = (-2.212 5.000
proj dx dy    = (0.000 0.000)


ordered pair 7: <14 9>
point 14 proj = (-10.302 5.000)
point  9 proj = (-10.302 5.000
proj dx dy    = (0.000 0.000)


ordered pair 8: <13 10>
point 13 proj = (-4.425 5.000)
point 10 proj = (-4.425 5.000
proj dx dy    = (0.000 0.000)
```

```
   ordered pair 9: <12 11>
   point 12 proj = (-14.425 5.000)
   point 11 proj = (-14.425 5.000
   proj dx dy    = (0.000 0.000)

   curve partition 1 is reflectional

== curve partition 2 test ==

CCPRF:

  FPA:
    ordered pair 0: <2 3>
    point c = (15.000 0.000)

    ordered pair 1: <1 4>
    point d = (17.939 0.955)

  CPARF:
    == midpoint testing ==

    ordered pair 0: <2 3>
    point m = (15.000 0.000)
    dist m to axis = 0.000

    ordered pair 1: <1 4>
    point m = (17.939 0.955)
    dist m to axis = 0.000

    ordered pair 2: <0 5>
    point m = (8.894 3.894)
    dist m to axis = 5.590

    curve partition 2 is not reflectional
```

```
== curve partition 3 test ==

CCPRF:

  FPA:
    ordered pair 0: <3 4>
    point c = (12.939 -4.045)

    ordered pair 1: <2 5>
    point d = (13.894 -1.106)

  CPARF:
    == midpoint testing ==

    ordered pair 0: <3 4>
    point m = (12.939 -4.045)
    dist m to axis = 0.000

    ordered pair 1: <2 5>
    point m = (13.894 -1.106)
    dist m to axis = 0.000

    ordered pair 2: <1 6>
    point m = (13.894 -1.106)
    dist m to axis = 0.000

    ordered pair 3: <0 7>
    point m = (3.894 -1.106)
    dist m to axis = 9.511

    curve partition 3 is not reflectional
```

```
== curve partition 4 test ==

CCPRF:

  FPA:
    ordered pair 0: <4 5>
    point c = (11.833 -5.151)

    ordered pair 1: <3 6>
    point d = (8.894 -6.106)

  CPARF:
    == midpoint testing ==

    ordered pair 0: <4 5>
    point m = (11.833 -5.151)
    dist m to axis = 0.000

    ordered pair 1: <3 6>
    point m = (8.894 -6.106)
    dist m to axis = 0.000

    ordered pair 2: <2 7>
    point m = (8.894 -6.106)
    dist m to axis = 0.000

    ordered pair 3: <1 8>
    point m = (8.894 3.894)
    dist m to axis = 9.511

    curve partition 4 is not reflectional
```

```
== curve partition 5 test ==

CCPRF:

  FPA:
    ordered pair 0: <5 6>
    point c = (7.788 -7.212)

    ordered pair 1: <4 7>
    point d = (6.833 -10.151)

  CPARF:
    == midpoint testing ==

    ordered pair 0: <5 6>
    point m = (7.788 -7.212)
    dist m to axis = 0.000

    ordered pair 1: <4 7>
    point m = (6.833 -10.151)
    dist m to axis = 0.000

    ordered pair 2: <3 8>
    point m = (3.894 -1.106)
    dist m to axis = 5.590

    curve partition 5 is not reflectional
```

```
== curve partition 6 test ==

CCPRF:

  FPA:
    ordered pair 0: <6 7>
    point c = (2.788 -12.212)

    ordered pair 1: <5 8>
    point d = (2.788 -2.212)

  CPARF:
    == midpoint testing ==

    ordered pair 0: <6 7>
    point m = (2.788 -12.212)
    dist m to axis = 0.000

    ordered pair 1: <5 8>
    point m = (2.788 -2.212)
    dist m to axis = 0.000

    ordered pair 2: <4 9>
    point m = (2.788 -8.090)
    dist m to axis = 0.000

    ordered pair 3: <3 10>
    point m = (2.788 0.000)
    dist m to axis = 0.000

    ordered pair 4: <2 11>
    point m = (2.788 0.000)
    dist m to axis = 0.000
```

```
ordered pair 5: <1 12>
point m = (2.788 10.000)
dist m to axis = 0.000


ordered pair 6: <0 13>
point m = (2.788 10.000)
dist m to axis = 0.000


ordered pair 7: <19 14>
point m = (2.788 18.090)
dist m to axis = 0.000


ordered pair 8: <18 15>
point m = (2.788 12.212)
dist m to axis = 0.000


ordered pair 9: <17 16>
point m = (2.788 22.212)
dist m to axis = 0.000


== projection testing ==

ordered pair 0: <6 7>
point  6 proj = (2.788 -12.212)
point  7 proj = (2.788 -12.212
proj dx dy    = (0.000 0.000)


ordered pair 1: <5 8>
point  5 proj = (2.788 -2.212)
point  8 proj = (2.788 -2.212
proj dx dy    = (0.000 0.000)


ordered pair 2: <4 9>
point  4 proj = (2.788 -8.090)
point  9 proj = (2.788 -8.090
proj dx dy    = (0.000 0.000)
```

```
ordered pair 3: <3 10>
point  3 proj = (2.788 0.000)
point 10 proj = (2.788 0.000
proj dx dy    = (0.000 0.000)


ordered pair 4: <2 11>
point  2 proj = (2.788 0.000)
point 11 proj = (2.788 0.000
proj dx dy    = (0.000 0.000)


ordered pair 5: <1 12>
point  1 proj = (2.788 10.000)
point 12 proj = (2.788 10.000
proj dx dy    = (0.000 0.000)


ordered pair 6: <0 13>
point  0 proj = (2.788 10.000)
point 13 proj = (2.788 10.000
proj dx dy    = (0.000 0.000)


ordered pair 7: <19 14>
point 19 proj = (2.788 18.090)
point 14 proj = (2.788 18.090
proj dx dy    = (0.000 0.000)


ordered pair 8: <18 15>
point 18 proj = (2.788 12.212)
point 15 proj = (2.788 12.212
proj dx dy    = (0.000 0.000)
```

```
    ordered pair 9: <17 16>
    point 17 proj = (2.788 22.212)
    point 16 proj = (2.788 22.212
    proj dx dy   = (0.000 0.000)

    curve partition 6 is reflectional

== curve partition 7 test ==

CCPRF:

  FPA:
    ordered pair 0: <7 8>
    point c = (-2.212 -7.212)

    ordered pair 1: <6 9>
    point d = (-1.257 -10.151)

  CPARF:
    == midpoint testing ==

    ordered pair 0: <7 8>
    point m = (-2.212 -7.212)
    dist m to axis = 0.000

    ordered pair 1: <6 9>
    point m = (-1.257 -10.151)
    dist m to axis = 0.000

    ordered pair 2: <5 10>
    point m = (1.682 -1.106)
    dist m to axis = 5.590

    curve partition 7 is not reflectional
```

```
== curve partition 8 test ==

CCPRF:

  FPA:
    ordered pair 0: <8 9>
    point c = (-6.257 -5.151)

    ordered pair 1: <7 10>
    point d = (-3.318 -6.106)

  CPARF:
    == midpoint testing ==

    ordered pair 0: <8 9>
    point m = (-6.257 -5.151)
    dist m to axis = 0.000

    ordered pair 1: <7 10>
    point m = (-3.318 -6.106)
    dist m to axis = 0.000

    ordered pair 2: <6 11>
    point m = (-3.318 -6.106)
    dist m to axis = 0.000

    ordered pair 3: <5 12>
    point m = (-3.318 3.894)
    dist m to axis = 9.511

    curve partition 8 is not reflectional
```

```
== curve partition 9 test ==

CCPRF:

  FPA:
    ordered pair 0: <9 10>
    point c = (-7.364 -4.045)

    ordered pair 1: <8 11>
    point d = (-8.318 -1.106)

  CPARF:
    == midpoint testing ==

    ordered pair 0: <9 10>
    point m = (-7.364 -4.045)
    dist m to axis = 0.000

    ordered pair 1: <8 11>
    point m = (-8.318 -1.106)
    dist m to axis = 0.000

    ordered pair 2: <7 12>
    point m = (-8.318 -1.106)
    dist m to axis = 0.000

    ordered pair 3: <6 13>
    point m = (1.682 -1.106)
    dist m to axis = 9.511

    curve partition 9 is not reflectional
```

```
>> testing 10 <IPT IPT> curve partitions <<

== curve partition 0 test ==

CCPRF:

  FPA:
    ordered pair 0: <0 0>
    point c = (10.000 10.000)

    ordered pair 1: <19 1>
    point d = (17.939 14.045)

  CPARF:
    == midpoint testing ==

    ordered pair 0: <0 0>
    point m = (10.000 10.000)
    dist m to axis = 0.000

    ordered pair 1: <19 1>
    point m = (17.939 14.045)
    dist m to axis = 0.000

    ordered pair 2: <18 2>
    point m = (13.894 6.106)
    dist m to axis = 5.237

    curve partition 0 is not reflectional
```

```
== curve partition 1 test ==

CCPRF:

  FPA:
    ordered pair 0: <1 1>
    point c = (20.000 10.000)

    ordered pair 1: <0 2>
    point d = (15.000 5.000)

  CPARF:
    == midpoint testing ==

    ordered pair 0: <1 1>
    point m = (20.000 10.000)
    dist m to axis = 0.000

    ordered pair 1: <0 2>
    point m = (15.000 5.000)
    dist m to axis = 0.000

    ordered pair 2: <19 3>
    point m = (12.939 9.045)
    dist m to axis = 4.318

    curve partition 1 is not reflectional
```

```
== curve partition 2 test ==

CCPRF:

  FPA:
    ordered pair 0: <2 2>
    point c = (20.000 0.000)

    ordered pair 1: <1 3>
    point d = (15.000 5.000)

  CPARF:
    == midpoint testing ==

    ordered pair 0: <2 2>
    point m = (20.000 0.000)
    dist m to axis = 0.000

    ordered pair 1: <1 3>
    point m = (15.000 5.000)
    dist m to axis = 0.000

    ordered pair 2: <0 4>
    point m = (12.939 0.955)
    dist m to axis = 4.318

    curve partition 2 is not reflectional
```

```
== curve partition 3 test ==

CCPRF:

  FPA:
    ordered pair 0: <3 3>
    point c = (10.000 0.000)

    ordered pair 1: <2 4>
    point d = (17.939 -4.045)

  CPARF:
    == midpoint testing ==

    ordered pair 0: <3 3>
    point m = (10.000 0.000)
    dist m to axis = 0.000

    ordered pair 1: <2 4>
    point m = (17.939 -4.045)
    dist m to axis = 0.000

    ordered pair 2: <1 5>
    point m = (13.894 3.894)
    dist m to axis = 5.237

    curve partition 3 is not reflectional
```

```
== curve partition 4 test ==

CCPRF:

  FPA:
    ordered pair 0: <4 4>
    point c = (15.878 -8.090)

    ordered pair 1: <3 5>
    point d = (8.894 -1.106)

  CPARF:
    == midpoint testing ==

    ordered pair 0: <4 4>
    point m = (15.878 -8.090)
    dist m to axis = 0.000

    ordered pair 1: <3 5>
    point m = (8.894 -1.106)
    dist m to axis = 0.000

    ordered pair 2: <2 6>
    point m = (13.894 -6.106)
    dist m to axis = 0.000

    ordered pair 3: <1 7>
    point m = (8.894 -1.106)
    dist m to axis = 0.000

    ordered pair 4: <0 8>
    point m = (3.894 3.894)
    dist m to axis = 0.000
```

```
ordered pair 5: <19 9>
point m = (2.788 5.000)
dist m to axis = 0.000


ordered pair 6: <18 10>
point m = (1.682 6.106)
dist m to axis = 0.000


ordered pair 7: <17 11>
point m = (-3.318 11.106)
dist m to axis = 0.000


ordered pair 8: <16 12>
point m = (-8.318 16.106)
dist m to axis = 0.000


ordered pair 9: <15 13>
point m = (-3.318 11.106)
dist m to axis = 0.000


ordered pair 10: <14 14>
point m = (-10.302 18.090)
dist m to axis = 0.000


== projection testing ==


ordered pair 0: <4 4>
point  4 proj = (15.878 -8.090)
point  4 proj = (15.878 -8.090
proj dx dy    = (0.000 0.000)


ordered pair 1: <3 5>
point  3 proj = (8.894 -1.106)
point  5 proj = (8.894 -1.106
proj dx dy    = (0.000 0.000)
```

```
ordered pair 2: <2 6>
point  2 proj = (13.894 -6.106)
point  6 proj = (13.894 -6.106
proj dx dy    = (0.000 0.000)


ordered pair 3: <1 7>
point  1 proj = (8.894 -1.106)
point  7 proj = (8.894 -1.106
proj dx dy    = (0.000 0.000)


ordered pair 4: <0 8>
point  0 proj = (3.894 3.894)
point  8 proj = (3.894 3.894
proj dx dy    = (0.000 0.000)


ordered pair 5: <19 9>
point 19 proj = (2.788 5.000)
point  9 proj = (2.788 5.000
proj dx dy    = (0.000 0.000)


ordered pair 6: <18 10>
point 18 proj = (1.682 6.106)
point 10 proj = (1.682 6.106
proj dx dy    = (0.000 0.000)


ordered pair 7: <17 11>
point 17 proj = (-3.318 11.106)
point 11 proj = (-3.318 11.106
proj dx dy    = (0.000 0.000)


ordered pair 8: <16 12>
point 16 proj = (-8.318 16.106)
point 12 proj = (-8.318 16.106
proj dx dy    = (0.000 0.000)
```

```
ordered pair 9: <15 13>
point 15 proj = (-3.318 11.106)
point 13 proj = (-3.318 11.106
proj dx dy    = (0.000 0.000)


ordered pair 10: <14 14>
point 14 proj = (-10.302 18.090)
point 14 proj = (-10.302 18.090
proj dx dy    = (0.000 0.000)


curve partition 4 is reflectional

== curve partition 5 test ==

CCPRF:

  FPA:
    ordered pair 0: <5 5>
    point c = (7.788 -2.212)


    ordered pair 1: <4 6>
    point d = (11.833 -10.151)


  CPARF:
    == midpoint testing ==

    ordered pair 0: <5 5>
    point m = (7.788 -2.212)
    dist m to axis = 0.000


    ordered pair 1: <4 6>
    point m = (11.833 -10.151)
    dist m to axis = 0.000
```

```
    ordered pair 2: <3 7>
    point m = (3.894 -6.106)
    dist m to axis = 5.237


    curve partition 5 is not reflectional


== curve partition 6 test ==

CCPRF:

  FPA:
    ordered pair 0: <6 6>
    point c = (7.788 -12.212)


    ordered pair 1: <5 7>
    point d = (2.788 -7.212)


  CPARF:
    == midpoint testing ==


    ordered pair 0: <6 6>
    point m = (7.788 -12.212)
    dist m to axis = 0.000


    ordered pair 1: <5 7>
    point m = (2.788 -7.212)
    dist m to axis = 0.000


    ordered pair 2: <4 8>
    point m = (6.833 -5.151)
    dist m to axis = 4.318


    curve partition 6 is not reflectional
```

```
== curve partition 7 test ==

CCPRF:

  FPA:
    ordered pair 0: <7 7>
    point c = (−2.212 −12.212)

    ordered pair 1: <6 8>
    point d = (2.788 −7.212)

  CPARF:
    == midpoint testing ==

    ordered pair 0: <7 7>
    point m = (−2.212 −12.212)
    dist m to axis = 0.000

    ordered pair 1: <6 8>
    point m = (2.788 −7.212)
    dist m to axis = 0.000

    ordered pair 2: <5 9>
    point m = (−1.257 −5.151)
    dist m to axis = 4.318

    curve partition 7 is not reflectional
```

```
== curve partition 8 test ==

CCPRF:

  FPA:
    ordered pair 0: <8 8>
    point c = (-2.212 -2.212)

    ordered pair 1: <7 9>
    point d = (-6.257 -10.151)

  CPARF:
    == midpoint testing ==

    ordered pair 0: <8 8>
    point m = (-2.212 -2.212)
    dist m to axis = 0.000

    ordered pair 1: <7 9>
    point m = (-6.257 -10.151)
    dist m to axis = 0.000

    ordered pair 2: <6 10>
    point m = (1.682 -6.106)
    dist m to axis = 5.237

    curve partition 8 is not reflectional
```

```
== curve partition 9 test ==

CCPRF:

  FPA:
    ordered pair 0: <9 9>
    point c = (-10.302 -8.090)

    ordered pair 1: <8 10>
    point d = (-3.318 -1.106)

  CPARF:
    == midpoint testing ==

    ordered pair 0: <9 9>
    point m = (-10.302 -8.090)
    dist m to axis = 0.000

    ordered pair 1: <8 10>
    point m = (-3.318 -1.106)
    dist m to axis = 0.000

    ordered pair 2: <7 11>
    point m = (-8.318 -6.106)
    dist m to axis = 0.000

    ordered pair 3: <6 12>
    point m = (-3.318 -1.106)
    dist m to axis = 0.000

    ordered pair 4: <5 13>
    point m = (1.682 3.894)
    dist m to axis = 0.000
```

```
ordered pair 5: <4 14>
point m = (2.788 5.000)
dist m to axis = 0.000

ordered pair 6: <3 15>
point m = (3.894 6.106)
dist m to axis = 0.000

ordered pair 7: <2 16>
point m = (8.894 11.106)
dist m to axis = 0.000

ordered pair 8: <1 17>
point m = (13.894 16.106)
dist m to axis = 0.000

ordered pair 9: <0 18>
point m = (8.894 11.106)
dist m to axis = 0.000

ordered pair 10: <19 19>
point m = (15.878 18.090)
dist m to axis = 0.000

== projection testing ==

ordered pair 0: <9 9>
point  9 proj = (-10.302 -8.090)
point  9 proj = (-10.302 -8.090)
proj dx dy    = (0.000 0.000)

ordered pair 1: <8 10>
point  8 proj = (-3.318 -1.106)
point 10 proj = (-3.318 -1.106
proj dx dy    = (0.000 0.000)
```

```
ordered pair 2: <7 11>
point  7 proj = (−8.318 −6.106)
point 11 proj = (−8.318 −6.106
proj dx dy   = (0.000 0.000)


ordered pair 3: <6 12>
point  6 proj = (−3.318 −1.106)
point 12 proj = (−3.318 −1.106
proj dx dy   = (0.000 0.000)


ordered pair 4: <5 13>
point  5 proj = (1.682 3.894)
point 13 proj = (1.682 3.894
proj dx dy   = (0.000 0.000)


ordered pair 5: <4 14>
point  4 proj = (2.788 5.000)
point 14 proj = (2.788 5.000
proj dx dy   = (0.000 0.000)


ordered pair 6: <3 15>
point  3 proj = (3.894 6.106)
point 15 proj = (3.894 6.106
proj dx dy   = (0.000 0.000)


ordered pair 7: <2 16>
point  2 proj = (8.894 11.106)
point 16 proj = (8.894 11.106
proj dx dy   = (0.000 0.000)


ordered pair 8: <1 17>
point  1 proj = (13.894 16.106)
point 17 proj = (13.894 16.106
proj dx dy   = (0.000 0.000)
```

```
ordered pair 9: <0 18>
point  0 proj = (8.894 11.106)
point 18 proj = (8.894 11.106
proj dx dy    = (0.000 0.000)


ordered pair 10: <19 19>
point 19 proj = (15.878 18.090)
point 19 proj = (15.878 18.090
proj dx dy    = (0.000 0.000)


curve partition 9 is reflectional

found 4 reflectional curve partitions
```

7.6.4 Discussion

The number of points in the AAC, 20, is even. Function DRFS therefore tests 10 <BLS BLS> and 10 <IPT IPT> curve partitions. DRFS begins testing <BLS BLS> curve partitions on page 155 and <IPT IPT> curve partitions on page 169.

DRFS determines that <BLS BLS> curve partitions 1 and 6 are reflectional, and that <BLS BLS> curve partitions 0, 2 – 5, and 7 – 9 are non-reflectional. <BLS BLS> curve partition 1 is $<< 1\ 2 >< 0\ 3 >< 19\ 4 >< 18\ 5 >< 17\ 6 >< 16\ 7 >< 15\ 8 >$ $< 14\ 9 >< 13\ 10 >< 12\ 11 >>$. <BLS BLS> curve partition 6 is $<< 6\ 7 >< 5\ 8 >$ $< 4\ 9 >< 3\ 10 >< 2\ 11 >< 1\ 12 >< 0\ 13 >< 19\ 14 >< 18\ 15 >< 17\ 16 >>$. The AAC's graph in Figure 7.14 on page 153 indicates that <BLS BLS> curve partitions 1 and 6 are indeed reflectional.

DRFS determines that <IPT IPT> curve partitions 4 and 9 are reflectional, and that <IPT IPT> curve partitions 0 – 3 and 5 – 8 are non-reflectional. <IPT IPT> curve partition 4 is $<< 4\ 4 >< 3\ 5 >< 2\ 6 >< 1\ 7 >< 0\ 8 >< 19\ 9 >< 18\ 10 >< 17\ 11 >$ $< 16\ 12 >< 15\ 13 >< 14\ 14 >>$. <IPT IPT> curve partition 9 is $<< 9\ 9 >< 8\ 10 >$ $< 7\ 11 >< 6\ 12 >< 5\ 13 >< 4\ 14 >< 3\ 15 >< 2\ 16 >< 1\ 17 >< 0\ 18 >< 19\ 19 >>$. The AAC's graph indicates that <IPT IPT> curve partitions 4 and 9 are indeed reflectional.

183

Each curve partition is tested by function CCPRF. CCPRF calls function FPA to generate a partition axis for the curve partition as described in section 7.2, and then calls function CPARF to check whether the partition axis is reflectional. CPARF performs midpoint and projection testing on the curve partition's ordered pairs as described in section 7.3.

DRFS completes its analysis by reporting that it found a total of 4 reflectional curve partitions among the 20 curve partitions tested.

7.7   DRFS example 2

This section provides an example of the DRFS algorithm operating on closed AAC [300: 0 160 240 240], which has an odd number of points. The example is divided into four parts: (1) a graph of the AAC; (2) a listing containing the AAC's point data; (3) a trace of DRFS algorithm operation; and (4) a discussion of the trace.

7.7.1   Graph

Figure 7.15 on the facing page shows that the AAC has three reflectional symmetries.

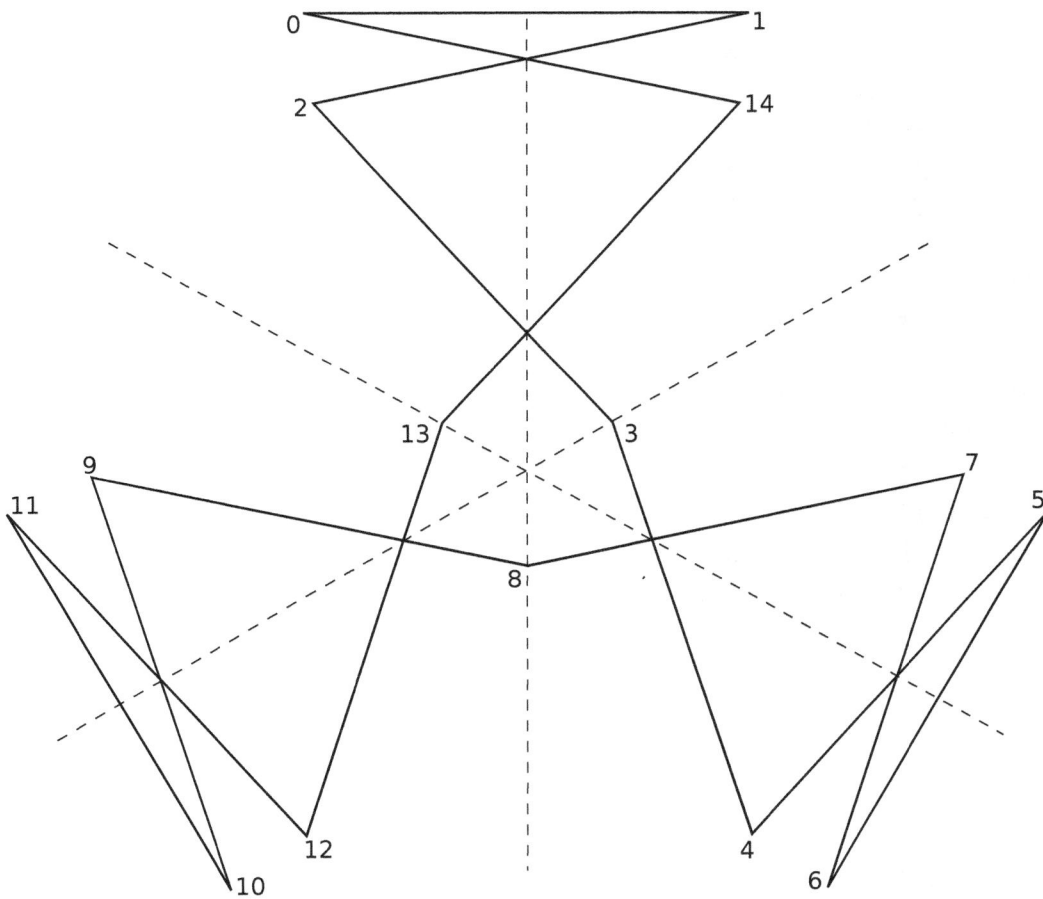

Figure 7.15   AAC [300: 0 160 240 240]

7.7.2 Point data

Listing 7.2 shows the point data for the AAC.

Listing 7.2

```
pt[ 0] = ( 10.000   10.000)
pt[ 1] = ( 20.000   10.000)
pt[ 2] = ( 10.219    7.921)
pt[ 3] = ( 16.910    0.489)
pt[ 4] = ( 20.000   -9.021)
pt[ 5] = ( 26.691   -1.590)
pt[ 6] = ( 21.691  -10.250)
pt[ 7] = ( 24.781   -0.739)
pt[ 8] = ( 15.000   -2.818)
pt[ 9] = (  5.219   -0.739)
pt[10] = (  8.309  -10.250)
pt[11] = (  3.309   -1.590)
pt[12] = ( 10.000   -9.021)
pt[13] = ( 13.090    0.489)
pt[14] = ( 19.781    7.921)
```

7.7.3 Trace

The DRFS algorithm determines the number of reflectional symmetries of the AAC as follows:

```
DRFS:
  >> testing 7 <BLS IPT> curve partitions <<

  == curve partition 0 test ==

  CCPRF:

    FPA:
      ordered pair 0: <0 1>
      point c = (15.000 10.000)

      ordered pair 1: <14 2>
      point d = (15.000 7.921)

    CPARF:
      == midpoint testing ==

      ordered pair 0: <0 1>
      point m = (15.000 10.000)
      dist m to axis = 0.000

      ordered pair 1: <14 2>
      point m = (15.000 7.921)
      dist m to axis = 0.000

      ordered pair 2: <13 3>
      point m = (15.000 0.489)
      dist m to axis = 0.000
```

```
ordered pair 3: <12 4>
point m = (15.000 -9.021)
dist m to axis = 0.000


ordered pair 4: <11 5>
point m = (15.000 -1.590)
dist m to axis = 0.000


ordered pair 5: <10 6>
point m = (15.000 -10.250)
dist m to axis = 0.000


ordered pair 6: <9 7>
point m = (15.000 -0.739)
dist m to axis = 0.000


ordered pair 7: <8 8>
point m = (15.000 -2.818)
dist m to axis = 0.000


== projection testing ==

ordered pair 0: <0 1>
point  0 proj = (15.000 10.000)
point  1 proj = (15.000 10.000
proj dx dy    = (0.000 0.000)


ordered pair 1: <14 2>
point 14 proj = (15.000 7.921)
point  2 proj = (15.000 7.921
proj dx dy    = (0.000 0.000)


ordered pair 2: <13 3>
point 13 proj = (15.000 0.489)
point  3 proj = (15.000 0.489
proj dx dy    = (0.000 0.000)
```

```
ordered pair 3: <12 4>
point 12 proj = (15.000 -9.021)
point  4 proj = (15.000 -9.021
proj dx dy    = (0.000 0.000)


ordered pair 4: <11 5>
point 11 proj = (15.000 -1.590)
point  5 proj = (15.000 -1.590
proj dx dy    = (0.000 0.000)


ordered pair 5: <10 6>
point 10 proj = (15.000 -10.250)
point  6 proj = (15.000 -10.250
proj dx dy    = (0.000 0.000)


ordered pair 6: <9 7>
point  9 proj = (15.000 -0.739)
point  7 proj = (15.000 -0.739
proj dx dy    = (0.000 0.000)


ordered pair 7: <8 8>
point  8 proj = (15.000 -2.818)
point  8 proj = (15.000 -2.818
proj dx dy    = (0.000 0.000)


curve partition 0 is reflectional
```

```
== curve partition 1 test ==

CCPRF:

  FPA:
    ordered pair 0: <1 2>
    point c = (15.109 8.960)

    ordered pair 1: <0 3>
    point d = (13.455 5.245)

  CPARF:
    == midpoint testing ==

    ordered pair 0: <1 2>
    point m = (15.109 8.960)
    dist m to axis = 0.000

    ordered pair 1: <0 3>
    point m = (13.455 5.245)
    dist m to axis = 0.000

    ordered pair 2: <14 4>
    point m = (19.891 -0.550)
    dist m to axis = 8.236

    curve partition 1 is not reflectional
```

```
== curve partition 2 test ==

CCPRF:

  FPA:
    ordered pair 0: <2 3>
    point c = (13.564 4.205)

    ordered pair 1: <1 4>
    point d = (20.000 0.489)

  CPARF:
    == midpoint testing ==

    ordered pair 0: <2 3>
    point m = (13.564 4.205)
    dist m to axis = 0.000

    ordered pair 1: <1 4>
    point m = (20.000 0.489)
    dist m to axis = 0.000

    ordered pair 2: <0 5>
    point m = (18.346 4.205)
    dist m to axis = 2.391

    curve partition 2 is not reflectional
```

```
== curve partition 3 test ==

CCPRF:

  FPA:
    ordered pair 0: <3 4>
    point c = (18.455 -4.266)

    ordered pair 1: <2 5>
    point d = (18.455 3.166)

  CPARF:
    == midpoint testing ==

    ordered pair 0: <3 4>
    point m = (18.455 -4.266)
    dist m to axis = 0.000

    ordered pair 1: <2 5>
    point m = (18.455 3.166)
    dist m to axis = 0.000

    ordered pair 2: <1 6>
    point m = (20.846 -0.125)
    dist m to axis = 2.391

    curve partition 3 is not reflectional
```

```
== curve partition 4 test ==

CCPRF:

  FPA:
    ordered pair 0: <4 5>
    point c = (23.346 −5.305)

    ordered pair 1: <3 6>
    point d = (19.301 −4.880)

  CPARF:
    == midpoint testing ==

    ordered pair 0: <4 5>
    point m = (23.346 −5.305)
    dist m to axis = 0.000

    ordered pair 1: <3 6>
    point m = (19.301 −4.880)
    dist m to axis = 0.000

    ordered pair 2: <2 7>
    point m = (17.500 3.591)
    dist m to axis = 8.236

    curve partition 4 is not reflectional
```

```
== curve partition 5 test ==

CCPRF:

  FPA:
    ordered pair 0: <5 6>
    point c = (24.191 –5.920)

    ordered pair 1: <4 7>
    point d = (22.391 –4.880)

    ordered pair 2: <3 8>
    point d = (15.955 –1.165)

  CPARF:
    == midpoint testing ==

    ordered pair 0: <5 6>
    point m = (24.191 –5.920)
    dist m to axis = 0.000

    ordered pair 1: <4 7>
    point m = (22.391 –4.880)
    dist m to axis = 0.000

    ordered pair 2: <3 8>
    point m = (15.955 –1.165)
    dist m to axis = 0.000

    ordered pair 3: <2 9>
    point m = (7.719 3.591)
    dist m to axis = 0.000

    ordered pair 4: <1 10>
    point m = (14.154 –0.125)
    dist m to axis = 0.000
```

```
ordered pair 5: <0 11>
point m = (6.654 4.205)
dist m to axis = 0.000


ordered pair 6: <14 12>
point m = (14.891 -0.550)
dist m to axis = 0.000


ordered pair 7: <13 13>
point m = (13.090 0.489)
dist m to axis = 0.000


== projection testing ==

ordered pair 0: <5 6>
point  5 proj = (24.191 -5.920)
point  6 proj = (24.191 -5.920)
proj dx dy    = (0.000 0.000)


ordered pair 1: <4 7>
point  4 proj = (22.391 -4.880)
point  7 proj = (22.391 -4.880)
proj dx dy    = (0.000 0.000)


ordered pair 2: <3 8>
point  3 proj = (15.955 -1.165)
point  8 proj = (15.955 -1.165)
proj dx dy    = (0.000 0.000)


ordered pair 3: <2 9>
point  2 proj = (7.719 3.591)
point  9 proj = (7.719 3.591
proj dx dy    = (0.000 0.000)
```

```
ordered pair 4: <1 10>
point  1 proj = (14.154 −0.125)
point 10 proj = (14.154 −0.125
proj dx dy    = (0.000 0.000)


ordered pair 5: <0 11>
point  0 proj = (6.654 4.205)
point 11 proj = (6.654 4.205
proj dx dy    = (0.000 0.000)


ordered pair 6: <14 12>
point 14 proj = (14.891 −0.550)
point 12 proj = (14.891 −0.550
proj dx dy    = (0.000 0.000)


ordered pair 7: <13 13>
point 13 proj = (13.090 0.489)
point 13 proj = (13.090 0.489
proj dx dy    = (0.000 0.000)


curve partition 5 is reflectional
```

```
== curve partition 6 test ==

CCPRF:

  FPA:
    ordered pair 0: <6 7>
    point c = (23.236 -5.495)

    ordered pair 1: <5 8>
    point d = (20.846 -2.204)

  CPARF:
    == midpoint testing ==

    ordered pair 0: <6 7>
    point m = (23.236 -5.495)
    dist m to axis = 0.000

    ordered pair 1: <5 8>
    point m = (20.846 -2.204)
    dist m to axis = 0.000

    ordered pair 2: <4 9>
    point m = (12.609 -4.880)
    dist m to axis = 8.236

    curve partition 6 is not reflectional
```

```
>> testing 8 <IPT BLS> curve partitions <<

== curve partition 0 test ==

CCPRF:

  FPA:
    ordered pair 0: <0 0>
    point c = (10.000 10.000)

    ordered pair 1: <14 1>
    point d = (19.891 8.960)

  CPARF:
    == midpoint testing ==

    ordered pair 0: <0 0>
    point m = (10.000 10.000)
    dist m to axis = 0.000

    ordered pair 1: <14 1>
    point m = (19.891 8.960)
    dist m to axis = 0.000

    ordered pair 2: <13 2>
    point m = (11.654 4.205)
    dist m to axis = 5.590

    curve partition 0 is not reflectional
```

```
== curve partition 1 test ==

CCPRF:

  FPA:
    ordered pair 0: <1 1>
    point c = (20.000 10.000)

    ordered pair 1: <0 2>
    point d = (10.109 8.960)

  CPARF:
    == midpoint testing ==

    ordered pair 0: <1 1>
    point m = (20.000 10.000)
    dist m to axis = 0.000

    ordered pair 1: <0 2>
    point m = (10.109 8.960)
    dist m to axis = 0.000

    ordered pair 2: <14 3>
    point m = (18.346 4.205)
    dist m to axis = 5.590

    curve partition 1 is not reflectional
```

```
== curve partition 2 test ==

CCPRF:

  FPA:
    ordered pair 0: <2 2>
    point c = (10.219 7.921)

    ordered pair 1: <1 3>
    point d = (18.455 5.245)

  CPARF:
    == midpoint testing ==

    ordered pair 0: <2 2>
    point m = (10.219 7.921)
    dist m to axis = 0.000

    ordered pair 1: <1 3>
    point m = (18.455 5.245)
    dist m to axis = 0.000

    ordered pair 2: <0 4>
    point m = (15.000 0.489)
    dist m to axis = 5.590

    curve partition 2 is not reflectional
```

```
== curve partition 3 test ==

CCPRF:

  FPA:
    ordered pair 0: <3 3>
    point c = (16.910 0.489)

    ordered pair 1: <2 4>
    point d = (15.109 −0.550)

    ordered pair 2: <1 5>
    point d = (23.346 4.205)

  CPARF:
    == midpoint testing ==

    ordered pair 0: <3 3>
    point m = (16.910 0.489)
    dist m to axis = 0.000

    ordered pair 1: <2 4>
    point m = (15.109 −0.550)
    dist m to axis = 0.000

    ordered pair 2: <1 5>
    point m = (23.346 4.205)
    dist m to axis = 0.000

    ordered pair 3: <0 6>
    point m = (15.846 −0.125)
    dist m to axis = 0.000

    ordered pair 4: <14 7>
    point m = (22.281 3.591)
    dist m to axis = 0.000
```

```
ordered pair 5: <13 8>
point m = (14.045 -1.165)
dist m to axis = 0.000


ordered pair 6: <12 9>
point m = (7.609 -4.880)
dist m to axis = 0.000


ordered pair 7: <11 10>
point m = (5.809 -5.920)
dist m to axis = 0.000


== projection testing ==

ordered pair 0: <3 3>
point  3 proj = (16.910 0.489)
point  3 proj = (16.910 0.489
proj dx dy    = (0.000 0.000)


ordered pair 1: <2 4>
point  2 proj = (15.109 -0.550)
point  4 proj = (15.109 -0.550
proj dx dy    = (0.000 0.000)


ordered pair 2: <1 5>
point  1 proj = (23.346 4.205)
point  5 proj = (23.346 4.205
proj dx dy    = (0.000 0.000)


ordered pair 3: <0 6>
point  0 proj = (15.846 -0.125)
point  6 proj = (15.846 -0.125
proj dx dy    = (0.000 0.000)
```

```
ordered pair 4: <14 7>
point 14 proj = (22.281 3.591)
point  7 proj = (22.281 3.591
proj dx dy    = (0.000 0.000)


ordered pair 5: <13 8>
point 13 proj = (14.045 −1.165)
point  8 proj = (14.045 −1.165
proj dx dy    = (0.000 0.000)


ordered pair 6: <12 9>
point 12 proj = (7.609 −4.880)
point  9 proj = (7.609 −4.880
proj dx dy    = (0.000 0.000)


ordered pair 7: <11 10>
point 11 proj = (5.809 −5.920)
point 10 proj = (5.809 −5.920
proj dx dy    = (0.000 0.000)


curve partition 3 is reflectional
```

```
== curve partition 4 test ==

CCPRF:

  FPA:
    ordered pair 0: <4 4>
    point c = (20.000 -9.021)

    ordered pair 1: <3 5>
    point d = (21.801 -0.550)

  CPARF:
    == midpoint testing ==

    ordered pair 0: <4 4>
    point m = (20.000 -9.021)
    dist m to axis = 0.000

    ordered pair 1: <3 5>
    point m = (21.801 -0.550)
    dist m to axis = 0.000

    ordered pair 2: <2 6>
    point m = (15.955 -1.165)
    dist m to axis = 5.590

    curve partition 4 is not reflectional
```

```
== curve partition 5 test ==

CCPRF:

  FPA:
    ordered pair 0: <5 5>
    point c = (26.691 -1.590)

    ordered pair 1: <4 6>
    point d = (20.846 -9.636)

  CPARF:
    == midpoint testing ==

    ordered pair 0: <5 5>
    point m = (26.691 -1.590)
    dist m to axis = 0.000

    ordered pair 1: <4 6>
    point m = (20.846 -9.636)
    dist m to axis = 0.000

    ordered pair 2: <3 7>
    point m = (20.846 -0.125)
    dist m to axis = 5.590

    curve partition 5 is not reflectional
```

```
== curve partition 6 test ==

CCPRF:

  FPA:
    ordered pair 0: <6 6>
    point c = (21.691 -10.250)

    ordered pair 1: <5 7>
    point d = (25.736 -1.165)

  CPARF:
    == midpoint testing ==

    ordered pair 0: <6 6>
    point m = (21.691 -10.250)
    dist m to axis = 0.000

    ordered pair 1: <5 7>
    point m = (25.736 -1.165)
    dist m to axis = 0.000

    ordered pair 2: <4 8>
    point m = (17.500 -5.920)
    dist m to axis = 5.590

    curve partition 6 is not reflectional
```

```
== curve partition 7 test ==

CCPRF:

  FPA:
    ordered pair 0: <7 7>
    point c = (24.781 -0.739)

    ordered pair 1: <6 8>
    point d = (18.346 -6.534)

  CPARF:
    == midpoint testing ==

    ordered pair 0: <7 7>
    point m = (24.781 -0.739)
    dist m to axis = 0.000

    ordered pair 1: <6 8>
    point m = (18.346 -6.534)
    dist m to axis = 0.000

    ordered pair 2: <5 9>
    point m = (15.955 -1.165)
    dist m to axis = 5.590

    curve partition 7 is not reflectional

 found 3 reflectional curve partitions
```

7.7.4 Discussion

The number of points in the AAC, 15, is odd. Function DRFS therefore tests 7
<BLS IPT> and 8 <IPT BLS> curve partitions. DRFS begins testing <BLS IPT> curve
partitions on page 187 and <IPT BLS> curve partitions on page 198.

DRFS determines that <BLS IPT> curve partitions 0 and 5 are reflectional, and
that <BLS IPT> curve partitions 1 – 4 and 6 are non-reflectional. <BLS IPT> curve
partition 0 is $<< 0\ 1 >< 14\ 2 >< 13\ 3 >< 12\ 4 >< 11\ 5 >< 10\ 6 >< 9\ 7 >< 8\ 8 >>$.
<BLS IPT> curve partition 5 is $<< 5\ 6 >< 4\ 7 >< 3\ 8 >< 2\ 9 >< 1\ 10 >< 0\ 11 >$
$< 14\ 12 >< 13\ 13 >>$. The AAC's graph in Figure 7.15 on page 185 indicates that
<BLS IPT> curve partitions 0 and 5 are indeed reflectional.

DRFS determines that <IPT BLS> curve partition 3 is reflectional, and that <IPT BLS>
curve partitions 0 – 2 and 4 – 7 are non-reflectional. <IPT BLS> curve partition 3
is $<< 3\ 3 >< 2\ 4 >< 1\ 5 >< 0\ 6 >< 14\ 7 >< 13\ 8 >< 12\ 9 >< 11\ 10 >>$. The AAC's
graph indicates that <IPT BLS> curve partition 3 is indeed reflectional.

DRFS completes its analysis by reporting that it found a total of 3 reflectional
curve partitions among the 15 curve partitions tested.

Notes

[1] This particular division between <BLS IPT> and <IPT BLS> curve partitions is arbitrary; a different DRFS algorithm design could employ a different division with no effect on the result. For example, the np curve partitions could be divided into np <BLS IPT> and zero <IPT BLS> curve partitions.

8 The Spirograph Connection

Spirograph® is a drafting tool that facilitates the drawing of two-dimensional closed curves called *hypotrochoids* and *epitrochoids* [6, p. 257]. This chapter proposes that for every order 1 accelerating angle curve (AAC1) there exists at least one corresponding Spirograph hypotrochoid curve (SHC) with the same rotational symmetry. The chapter begins with an introduction to SHC generation, and then describes the correspondence between AAC1 generation and SHC generation. Finally, three example AAC1s and their corresponding SHCs are provided.

8.1 SHC generation

This section describes the Spirograph ring/wheel assembly used to generate a SHC, the SHC signature, and how the hole number employed affects the SHC generated.

8.1.1 Spirograph ring/wheel assembly

Figure 8.1 on the following page shows two Spirograph gears assembled for hypotrochoid drawing. The inner gear, called a *wheel*, rolls around inside a stationary outer gear, called a *ring*. A pen is inserted into one of the *holes* in the wheel, and moves with the wheel as it rolls around the ring to draw a SHC. The location of the pen is represented by the blackened hole in the figure.

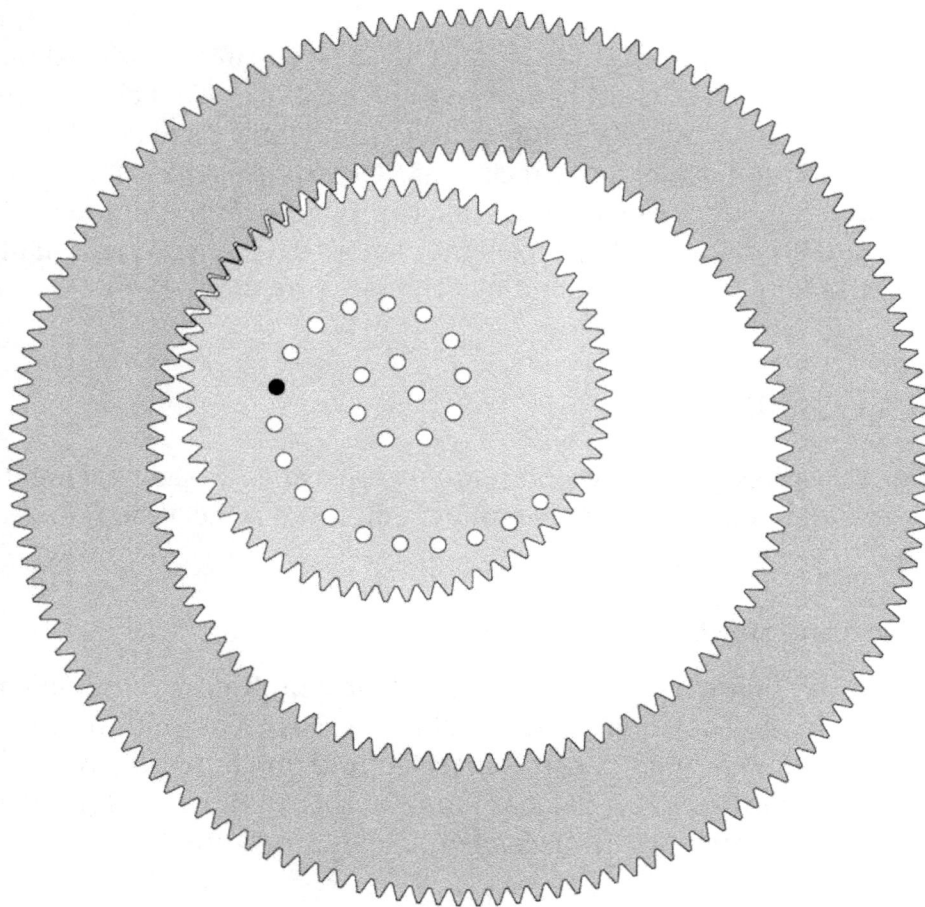

Figure 8.1   Spirograph ring/wheel assembly

A ring is identified by the number of teeth on its outer and inner edges, and a wheel is identified by the number of teeth on its outer edge. A typical Spirograph kit supplies $\frac{144}{96}$ and $\frac{150}{105}$ rings and 15 wheels ranging in size from 24 to 84 teeth. The holes in a wheel are numbered based on their distance from the wheel center, with hole 1 farthest from the center (i.e., closest to the wheel edge), and the other holes numbered in sequence as they approach the center. Figure 8.1 shows a 63-tooth wheel mated with a $\frac{144}{96}$ ring; the pen is inserted in hole 11.

8.1.2 SHC signature

The signature for the hypotrochoid curve generated by a Spirograph ring/wheel assembly has the form $[r : a_0\ w : h]$, where r is the number of teeth on the ring inner edge, a_0 is angular position of the ring/wheel assembly, w is the number of teeth on the wheel edge, and the h is the hole number.

The ring/wheel assembly angular position employs the geometric center of the ring as the center of rotation. Typically, when the ring/wheel assembly is used to generate a SHC the angular position of the assembly is not measured, so no numerical value is assigned to this parameter in the SHC signatures appearing in this chapter.

8.1.3 Effect of hole number on the SHC generated

Figures 8.2 on the next page, 8.3 on page 215, and 8.4 on page 216 demonstrate that changing the hole number can radically change the SHC generated by a Spirograph ring/wheel assembly. All three SHCs were generated by a 63-tooth wheel mated with a $\frac{144}{96}$ ring. The hole numbers employed were 1, 11, and 18, respectively.

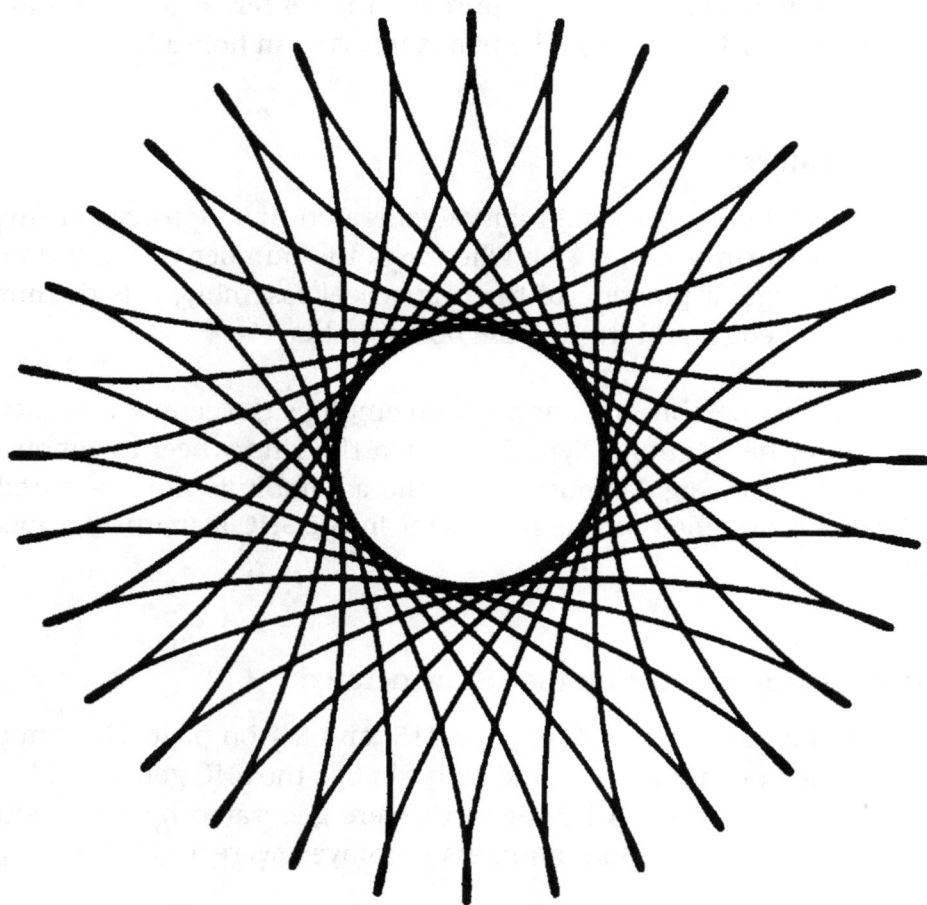

Figure 8.2 SHC [96: a_0 63: 1]

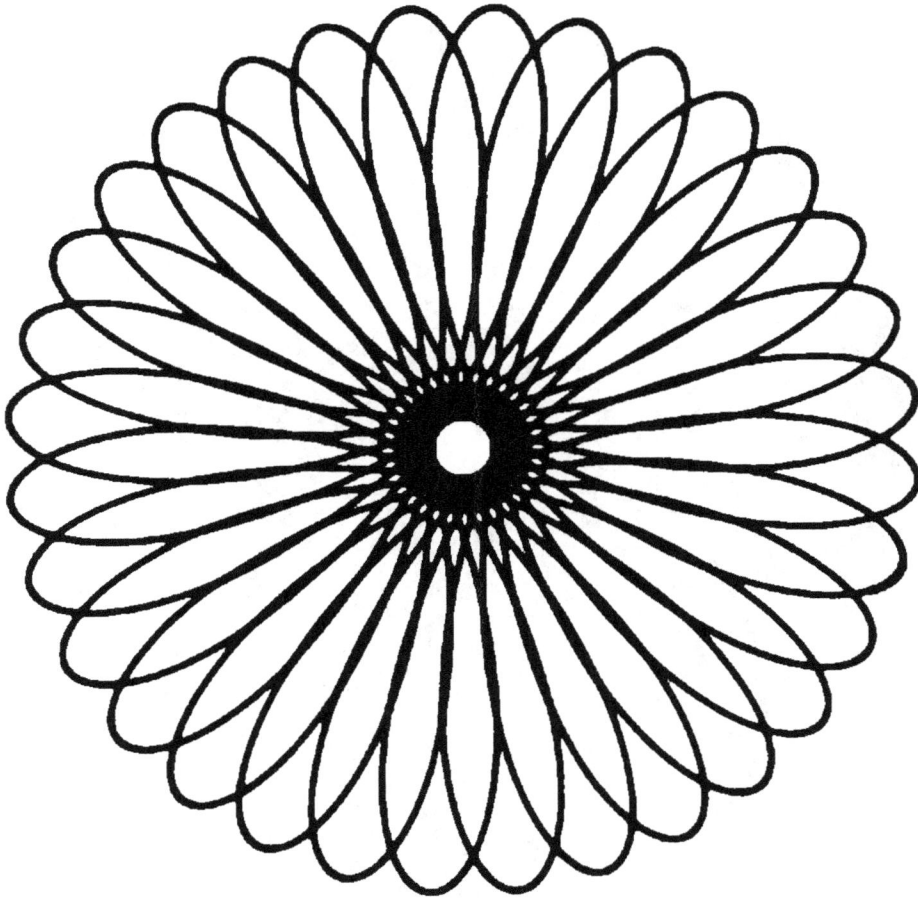

Figure 8.3 SHC [96: a_0 63: 11]

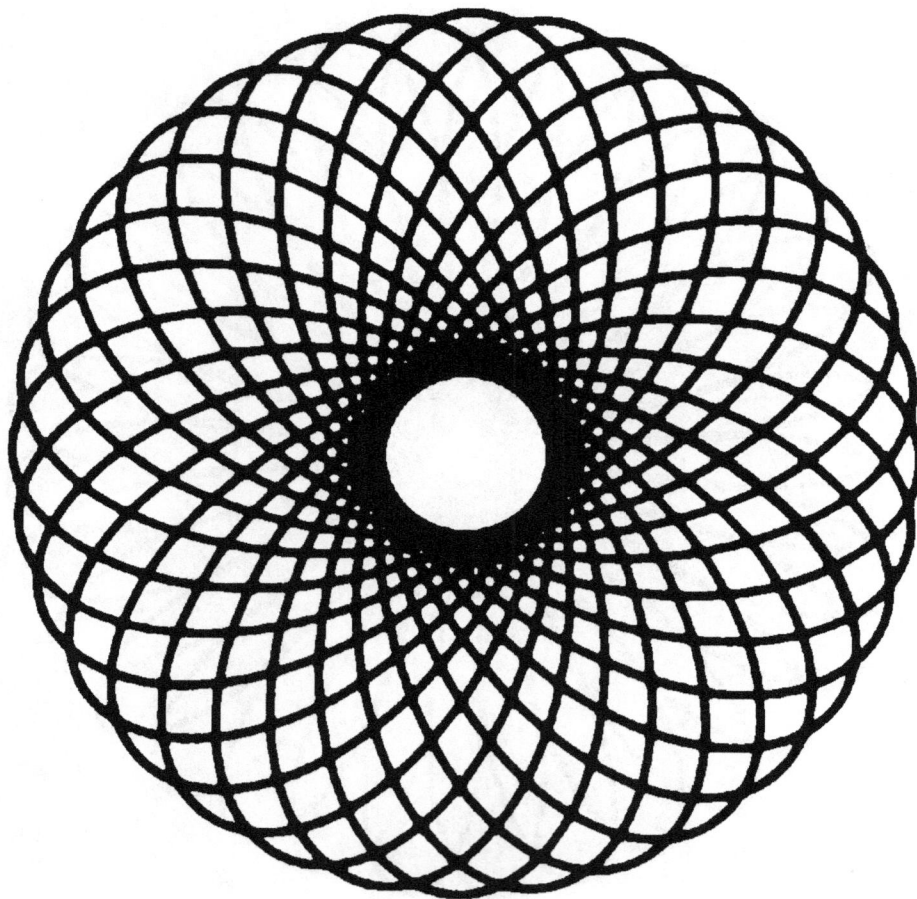

Figure 8.4 SHC [96: a_0 63: 18]

8.2 AAC1/SHC correspondence

Table 8.1 shows the correspondence between AAC1 generation parameters (GPs) and SHC generation parameters.

Table 8.1 AAC1/SHC generation parameter correspondence

AAC1 GP	SHC GP
line segment length, L	ring radius, R
number of slices per circle, A	number of teeth, inside of ring, r
initial base angle, a_0	angular position, ring/wheel assembly, a_0
initial first-order difference angle, a_1	number of teeth, wheel edge, w
N/A	hole number, h

The AAC1/SHC generation parameter correspondence is reflected in the design of the SHC signature. The AAC1 signature has the form $[A : a_0\ a_1]$ and the SHC signature has the form $[r : a_0\ w : h]$. The SHC signature format emphasizes the correspondence of A with r and a_1 with w. There is no AAC1 generation parameter corresponding to the hole number, h. The ring radius R is omitted from the SHC signature for the same reason that the line segment length L is omitted from the AAC1 signature: this parameter only determines the overall size of the curve and thus has no effect on its symmetry.

For an order 1 AAC, the order N AAC rotational symmetry prediction function PRTS given in Figure 5.1 on page 88 reduces to:

$$prts\_aac1 = \frac{\text{LCM}(A,a_1)}{a_1}$$

Given the correspondence between AAC1 and SHC generation parameters, the predicted rotational symmetry of a SHC is:

$$prts\_shc = \frac{\text{LCM}(r,w)}{w}$$

The next three sections provide examples of AAC1s and their corresponding SHCs.

8.3   AAC1/SHC example 1

Figure 8.5 on the facing page shows the AAC1 generated when the number of slices per circle is 96 and the base angle a_0 advances by 24 slices each iteration. Each iteration generates a line segment of the curve. The predicted rotational symmetry of the AAC1 is:

$$prts\_aac1 = \frac{\text{LCM}(A,a_1)}{a_1} = \frac{\text{LCM}(96,24)}{24} = 4$$

Figure 8.6 on page 220 shows the SHC generated when the ring inner edge has 96 teeth, the wheel edge has 24 teeth, and the pen is inserted in hole number 1. Each complete revolution of the wheel advances the wheel by 24 teeth along the inner edge of the ring. For each wheel revolution where drawing begins and ends with the hole located closest to the ring inner edge, a SHC arc analogous to an AAC1 line segment is generated. The predicted rotational symmetry of the SHC is:

$$prts\_shc = \frac{\text{LCM}(r,w)}{w} = \frac{\text{LCM}(96,24)}{24} = 4$$

Visual inspection of the AAC1 and SHC indicates that the predicted rotational symmetries are correct.

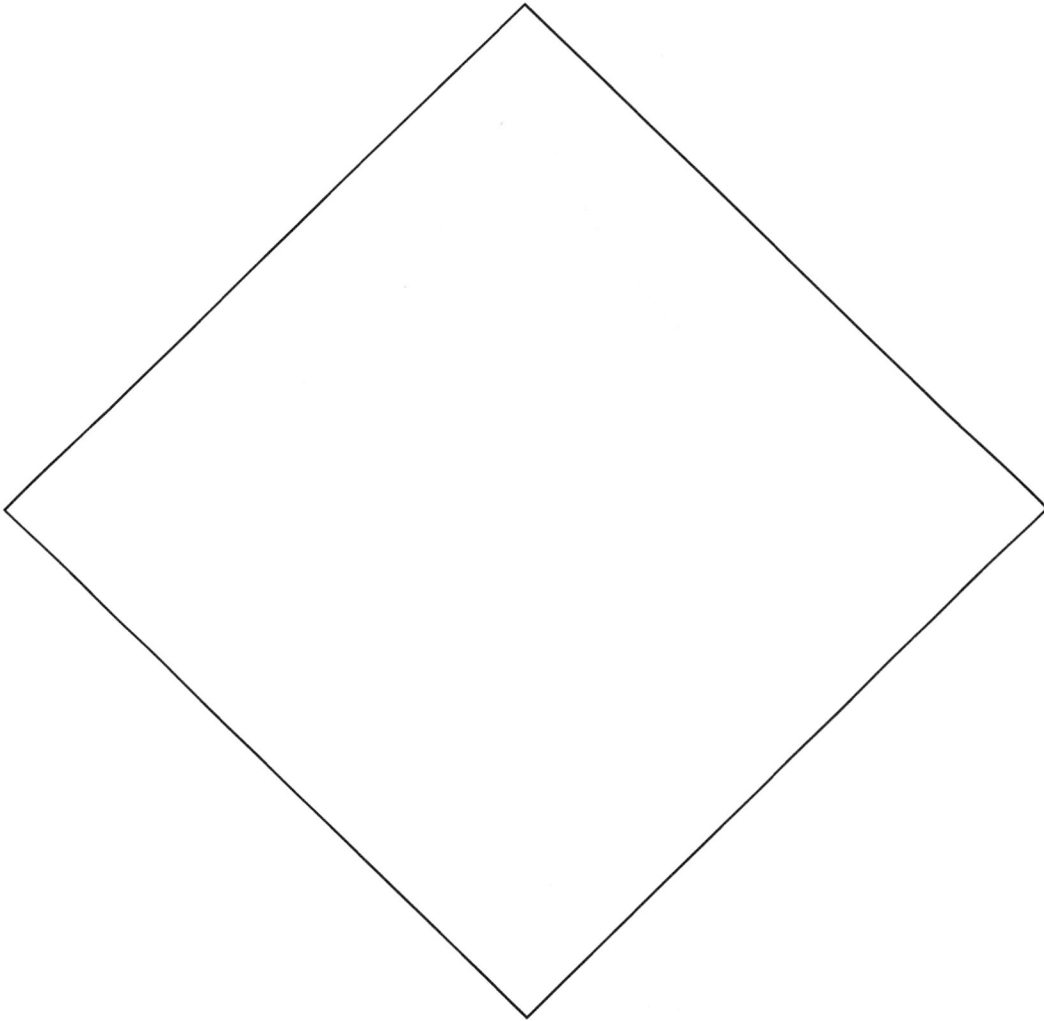

Figure 8.5   AAC [96: 12 24]

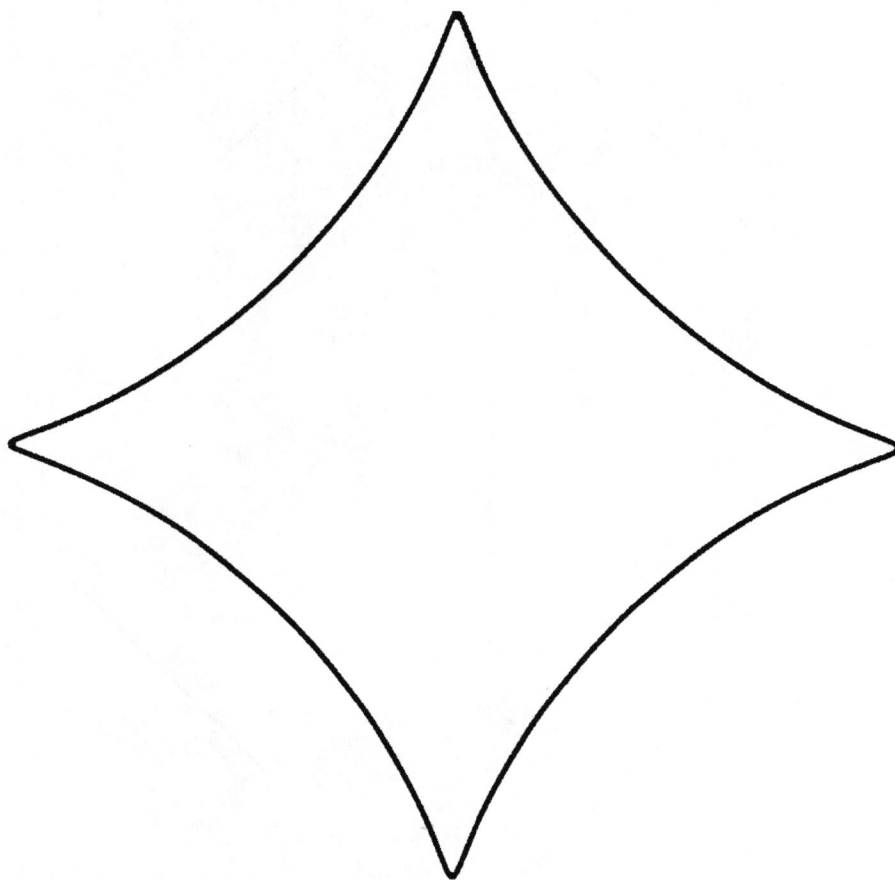

Figure 8.6   SHC [96: a_0 24: 1]

8.4 AAC1/SHC example 2

Figure 8.7 on the following page shows the AAC1 generated when the number of slices per circle is 105 and the base angle a_0 advances by 45 slices each iteration. Each iteration generates a line segment of the curve. The predicted rotational symmetry of the AAC1 is:

$$prts\_aac1 = \frac{\text{LCM}(A,a_1)}{a_1} = \frac{\text{LCM}(105,45)}{45} = 7$$

Figure 8.8 on page 223 shows the SHC generated when the ring inner edge has 105 teeth, the wheel edge has 45 teeth, and the pen is inserted in hole number 5. Each complete revolution of the wheel advances the wheel by 45 teeth along the inner edge of the ring. For each wheel revolution where drawing begins and ends with the hole located closest to the ring inner edge, a SHC arc analogous to an AAC1 line segment is generated. The predicted rotational symmetry of the SHC is:

$$prts\_shc = \frac{\text{LCM}(r,w)}{w} = \frac{\text{LCM}(105,45)}{45} = 7$$

Visual inspection of the AAC1 and SHC indicates that the predicted rotational symmetries are correct.

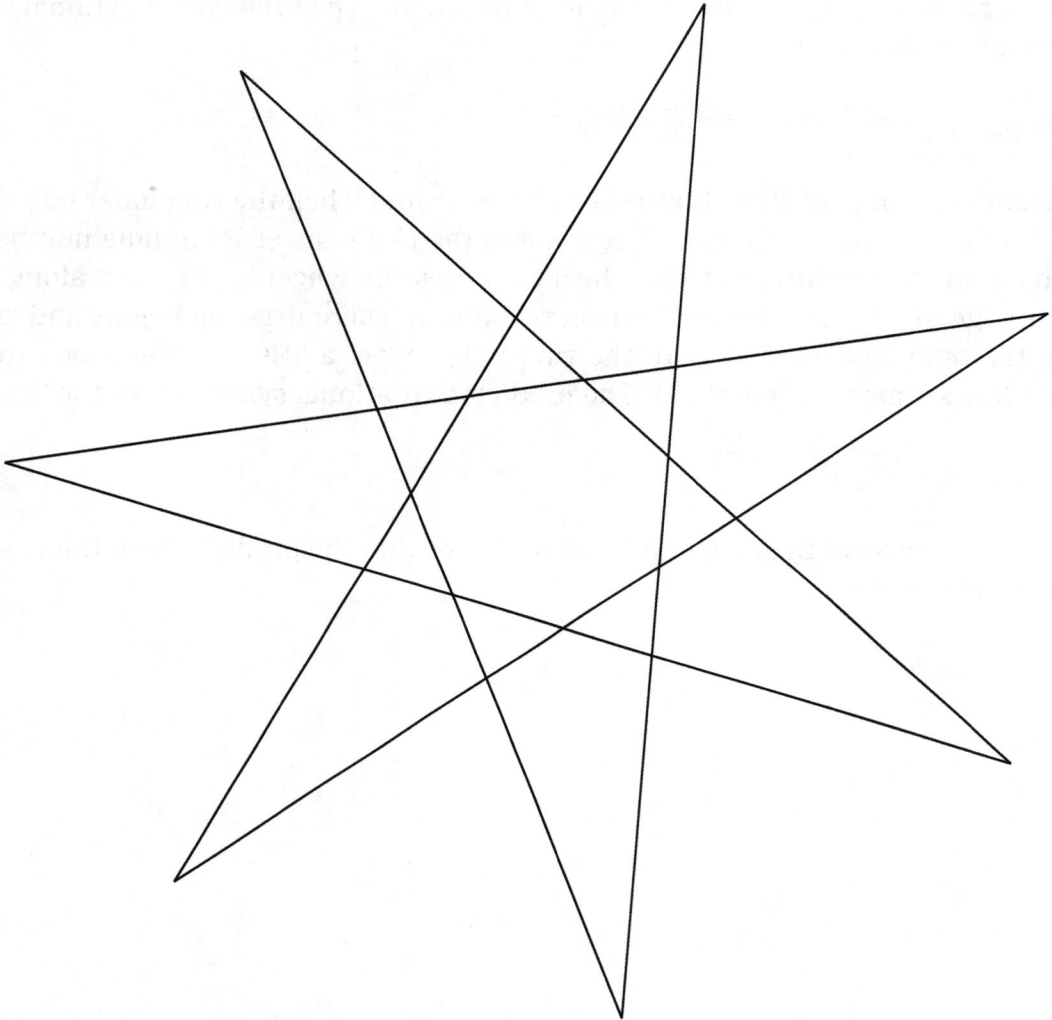

Figure 8.7   AAC [105: 10 45]

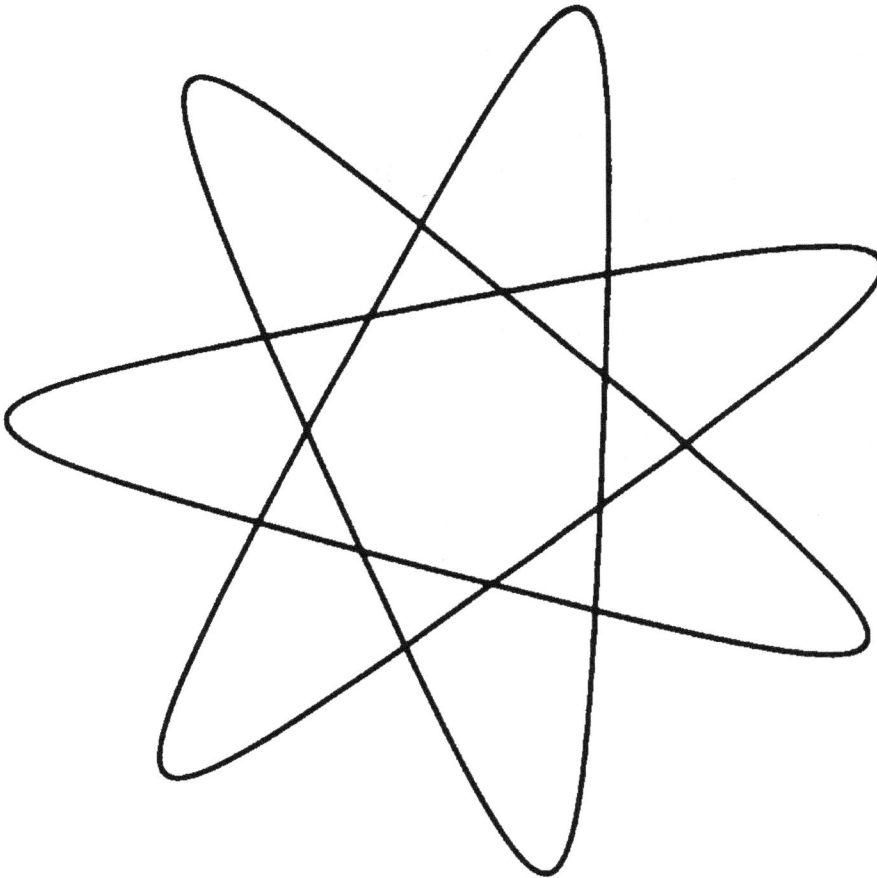

Figure 8.8   SHC [105: a_0 45: 5]

8.5 AAC1/SHC example 3

Figure 8.9 on the facing page shows the AAC1 generated when the number of slices per circle is 96 and the base angle a_0 advances by 56 slices each iteration. Each iteration generates a line segment of the curve. The predicted rotational symmetry of the AAC1 is:

$$prts\_aac1 = \frac{\text{LCM}_{(A,a_1)}}{a_1} = \frac{\text{LCM}_{(96,56)}}{56} = 12$$

Figure 8.10 on page 226 shows the SHC generated when the ring inner edge has 96 teeth, the wheel edge has 56 teeth, and the pen is inserted in hole number 10. Each complete revolution of the wheel advances the wheel by 56 teeth along the inner edge of the ring. For each wheel revolution where drawing begins and ends with the hole located closest to the ring inner edge, a SHC arc analogous to an AAC1 line segment is generated. The predicted rotational symmetry of the SHC is:

$$prts\_shc = \frac{\text{LCM}_{(r,w)}}{w} = \frac{\text{LCM}_{(96,56)}}{56} = 12$$

Visual inspection of the AAC1 and SHC indicates that the predicted rotational symmetries are correct.

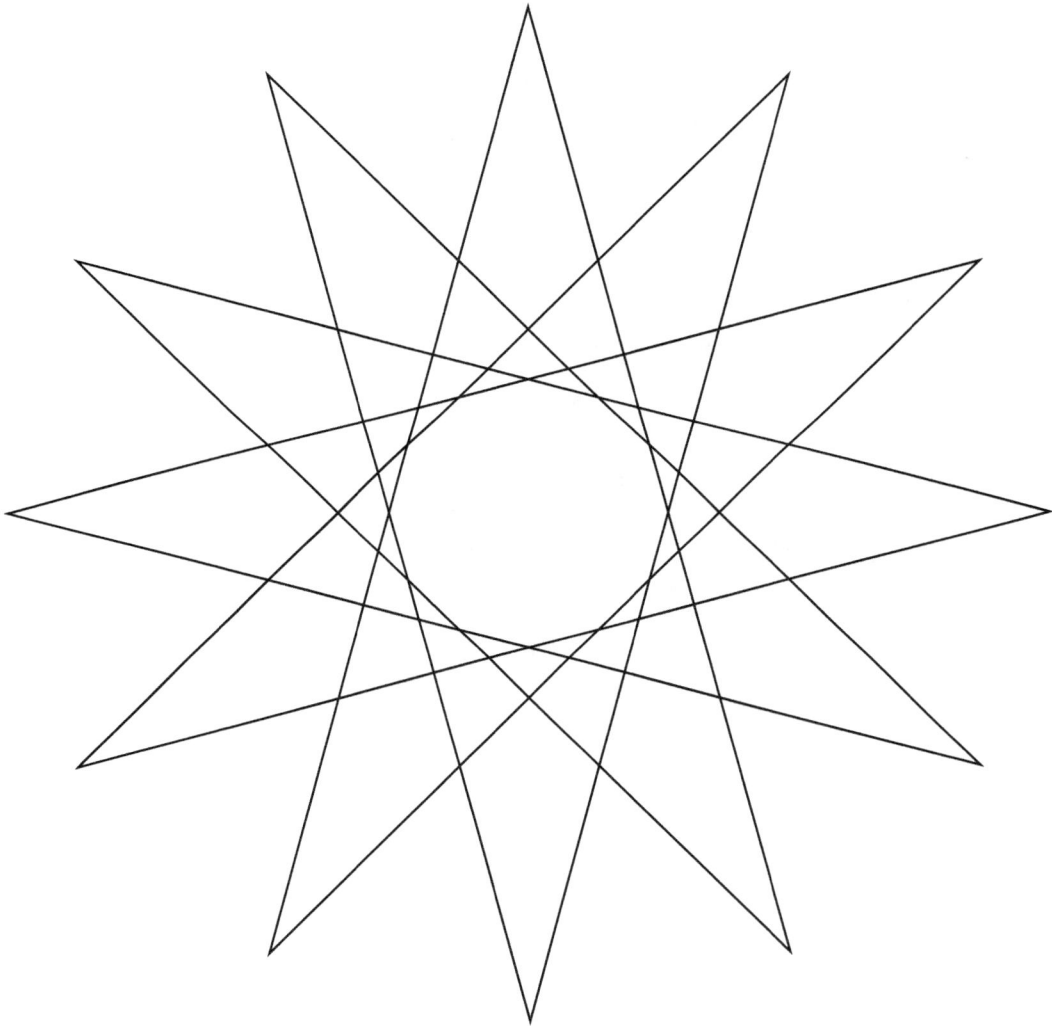

Figure 8.9 AAC [96: 4 56]

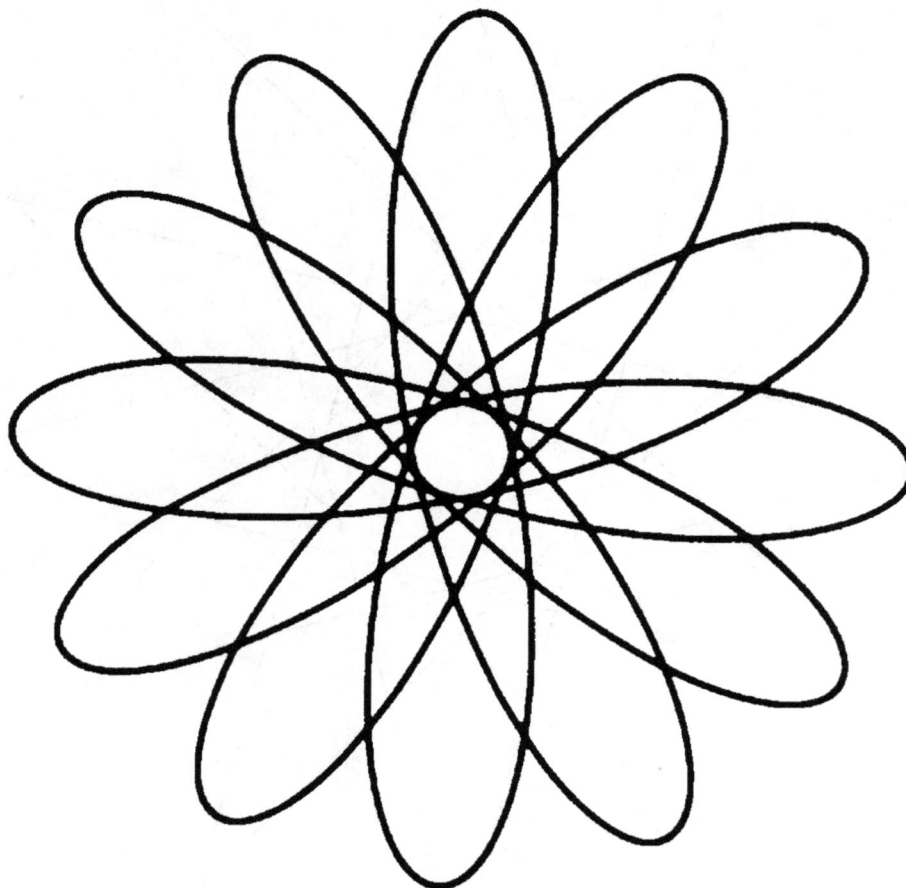

Figure 8.10   SHC [96: a_0 56: 10]

9 Open Questions

This chapter provides 15 open questions regarding AACs. The questions are organized into four categories: aesthetics, symmetry, Spirograph, and miscellaneous. Some questions have additional commentary to provide context and stimulate thought.

9.1 AAC aesthetics questions

Question 9.1. It is tedious to manually search through AACs to find those that are aesthetically pleasing. Can a machine learn a person's aesthetic preferences and then search for AACs aligned with those preferences?

Question 9.2. Are AACs sufficiently distinctive that a classifier can be trained to reliably discriminate between AACs and other types of curve?

9.2 AAC symmetry questions

Question 9.3. This book is primarily concerned with closed AACs because these tend to be more aesthetically pleasing than open AACs. However, in the interest of completeness the symmetry of open AACs should be analyzed. The analysis should include the prediction and determination of an open AAC's translational symmetry [8, p. 47].

Question 9.4. Given only an AAC's signature, how many reflectional symmetries does it have?

The PRTS algorithm presented in Chapter 5 predicts the number of rotational symmetries of an AAC given only its signature, and the DRTS algorithm presented in Chapter 6 determines the number of rotational symmetries of an AAC given only its point data. The DRTS algorithm provides an automatic method for checking the performance of the PRTS algorithm: the predicted number of rotational symmetries *before* AAC generation should equal the number of rotational symmetries determined *after* AAC generation.

The DRFS algorithm presented in Chapter 7 determines the number of reflectional symmetries of an AAC given only its point data. In the interest of completeness, an algorithm PRFS that predicts the number of reflectional symmetries of an AAC given only its signature should be developed. AAC reflectional symmetry analysis would then have a pair of algorithms (PRFS, DRFS) in the same way that AAC rotational symmetry analysis has a pair of algorithms (PRTS, DRTS).

9.3   Spirograph questions

Question 9.5. Is it possible to instantiate a physical Spirograph drafting tool that produces SHCs corresponding to order > 1 AACs?

If the answer is yes, then the physical instantiation may involve some kind of clever mechanical nesting of rolling wheels within rolling wheels. The best approach to this problem may be to first create a software-based Spirograph simulator generating Spirograph curves corresponding to higher-order AACs.[1] After thoroughly understanding the nature of such curves, the physical Spirograph instantiation problem may be easier to solve.

Question 9.6. A typical Spirograph kit provides internal gear shapes other than circular in order to expand the range of curve types that can be generated. For example, elliptical and triangular internal gear shapes may be provided. Interestingly, some of the curves generated by these non-circular internal gears are reminiscent of some higher-order accelerating angle curves. Do non-circular internal gears incorporate a type of acceleration into Spirograph curve generation that has a readily identifiable accelerating angle curve generation counterpart?

Question 9.7. For every order 1 AAC $[A : a_0 \ a_1]$ does there exist a corresponding SHC $[r : a_0 \ w : h]$ such that the SHC is *identical* to the AAC?

If the answer is no, then SHC generation is not a *perfect* generalization of order 1 AAC generation even though SHC generation has a parameter, hole number h, for which there is no corresponding AAC generation parameter. A key difference between order 1 AACs and SHCs is that the former are constructed from line segments whereas the latter are continuously varying.

This question can also be considered for a generalized Spirograph tool that also supports *hypocycloid* curve generation. A hypocycloid curve is generated when the pen is located at the wheel's edge. This capability is difficult to instantiate in a physical Spirograph drafting tool because a hole located at the wheel's edge would interfere with wheel/ring gear meshing. There are, however, software-based Spirograph simulators that can generate hypocycloid curves (see, for example, [5]).

Question 9.8. Is it possible to generalize order 1 AAC generation to include a parameter that has the same effect on the AACs generated as hole number has on the SHCs generated?

9.4 Miscellaneous questions

Question 9.9. Given only an AAC's signature, is it possible to predict whether the AAC is open or closed?

This problem is already partially solved: when algorithm PRTS described in Chapter 5 predicts that the AAC has nontrivial rotational symmetry, then it also necessarily predicts that the AAC is closed. Otherwise, the AAC may be either open or closed.

Question 9.10. Given some number of slices per circle A, are there an infinite number of *unique* AACs?

The answer to this question is 'no' if there is an AAC order N_{max} beyond which all AACs generated are identical to AACs with order $\leq N_{max}$.

Question 9.11. Given that AAC $[A : a_0, a_1, \cdots, a_N]$ generates some AAC c, can anything be reliably predicted about the nature of AAC c' generated by $[A : a'_0, a'_1, \cdots, a'_N]$, where $a'_i = A - a_i$?

This question asks what happens to an AAC when its initial base and difference angles are negated mod A. For example, if $A = 360$ and $a_i = 90$, then $A - a_i = 270$, which is equal to -90 mod 360.

Question 9.12. Predict how many AACs with number of slices per circle A and AAC order N:
(a) are identical to other AACs with the same A and N;
(b) are mirror images of other AACs with the same A and N;
(c) are closed (or open);
(d) have an even (or odd) number of lines.

Question 9.13. Given only an AAC's point data, what is the algorithm to generate the AAC's signature (or set of consistent signatures)?

Question 9.14. Is there a three-dimensional version of the AAC generation algorithm?

Question 9.15. Do the algorithms that determine the number of rotational and reflectional symmetries of an AAC given only the AAC's point data provided in chapters 6 and 7, respectively, return correct results for *any* closed, segmented, planar curve?

Notes

[1] Given the primitive state of computer and display technology in the early 1960s, creating a software-based Spirograph simulator was impractical for British engineer Denys Fisher, who developed the original Spirograph drafting tool [6, p. 255].

10 Conclusion

I have become an avid symmetry fan, addicted beyond cure, utterly convinced of the fertility of symmetry in scientific study and research as a unifying, clarifying, and simplifying factor. Moreover, far from being painful, these severe symptoms afford much pleasure, as I find in those aspects of symmetry with which I am concerned an [aesthetic] enjoyment ...

— Joe Rosen, *Symmetry Discovered* (1975)

This book has hopefully stimulated interest in the aesthetics and analysis of accelerating angle curves (AACs). The AAC generation algorithm is simple, but nevertheless can produce curves of remarkable complexity and elegance. The algorithms for predicting the number of rotational symmetries of a closed AAC given only its generation parameters and for determining the number of rotational and reflectional symmetries of a closed AAC given only its point data are straightforward. AACs and Spirograph curves are mutually complementary — by comparing the two curve generation methods, the understanding of both is enhanced.

There is much more aesthetic and analytical territory available for exploration by the interested reader. For example, as of this writing the author has generated AACs with number of slices per circle, A, of 3, 4, 5, 6, 8, 9, 10, 12, 20, 21, 35, 36, 45, 60, 80, 96, 105, 108, 120, 144, 240, 244, 248, 256, 336, 360, 361, 364, 378, 576, 666, 720, and 840. AACs with other values of A are *terra incognita*. Even for the values of A shown, only a few AACs have been generated within a vast (and possibly infinite) space of possible AACs. The Open Questions chapter provides suggestions for additional AAC analytical work; however, the most interesting AAC questions may as yet be unasked.

List of Abbreviations

AAC	Accelerating Angle Curve
AAC1	Accelerating Angle Curve, Order 1
AACG	Accelerating Angle Curve Generation
ALU	Arbitrary Length Unit
BLS	Bisect Line Segment
CCPRF	Check Curve Partition Reflectional
CNTRD	Centroid
CPARF	Check Partition Axis Reflectional
CRTS	Check Rotation Symmetric
DRFS	Determine [Number of] Reflectional Symmetries
DRTS	Determine [Number of] Rotational Symmetries
EFDM	Extrapolate Forward Differences [with] Modulo
FPA	Find Partition Axis
GP	Generation Parameter
IPT	Intercept Point
LCM	Least Common Multiple
PRFS	Predict [Number of] Reflectional Symmetries
PRTS	Predict [Number of] Rotational Symmetries
SHC	Spirograph Hypotrochoid Curve

Variable names employed in algorithm pseudocode have been excluded from this list.

Acknowledgements

The following tools and technologies were employed in the development of this book:

- **Xcode** : Macintosh application used for software development

- **Mike's Arbitrary Precision Math Library** : used by the AAC rotational symmetry prediction software

- **Mersenne Twister pseudorandom number generator** : used when pseudorandomly generating AACs

- **Scalable Vector Graphics** : graphics specification used for expressing AACs

- **Inkscape** : Macintosh application used to display and annotate AACs

- **Inspirograph** : software simulation of a Spirograph drafting tool [2].

The author wishes to thank the individuals who reviewed the content of this book prior to publication.

Bibliography

[1] Conway, J., Burgiel, H., & Goodman-Strauss, C. (2008). *The Symmetries of Things*. New York: CRC Press.

[2] Friend, N. (2016, May 10). *Inspirograph*.
www.nathanfriend.io/inspirograph

[3] Kim, S. (1981). *Inversions*. Peterborough, NH: McGraw-Hill.

[4] Lockwood, E.H. & Macmillan, R.H. (1978). *Geometric Symmetry*. New York: Cambridge University Press.

[5] Ortiz, M. (2014, January 23). *Spirograph Simulator*.
www.geogebra.org/m/ukXWe56e

[6] Posamentier, A.S. & Geretschläger, R. (2016). *The Circle: A Mathematical Exploration Beyond the Line*. New York: Prometheus Books.

[7] Rosen, J. (1998). *Symmetry Discovered: Concepts and Applications in Nature and Science*. Mineola, NY: Dover Publications.

[8] Weyl, H. (1952). *Symmetry*. Princeton, NJ: Princeton University Press.

Index

www.ingramcontent.com/pod-product-compliance
Lightning Source LLC
Chambersburg PA
CBHW051208200326
41519CB00025B/7045